Beyond Provenance
New Approaches to Interpreting the Chemistry of Archaeological Copper Alloys

Studies in Archaeological Sciences 6

The series Studies in Archaeological Sciences presents state-of-the-art methodological, technical or material science contributions to Archaeological Sciences. The series aims to reconstruct the integrated story of human and material culture through time and testifies to the necessity of inter- and multidisciplinary research in cultural heritage studies.

Beyond Provenance

New Approaches to Interpreting the Chemistry of Archaeological Copper Alloys

A.M. Pollard
With
P. Bray, A. Cuénod, P. Hommel, Y.-K. Hsu, R. Liu,
L. Perucchetti, J. Pouncett and M. Saunders

Leuven University Press

Published with support of

ISBN 978 94 6270 162 5
eISBN 978 94 6166 266 8

D / 2018/ 1869 / 44
NUR: 682

Lay-out: Friedemann Vervoort
Cover: Jurgen Leemans

Preface: FLAME and the 'Oxford System'

The intention of this volume is to present a coherent explanation of what has become known as 'the Oxford System' for interpreting the chemical and isotopic data from archaeological copper alloy objects. This system began with the DPhil of Peter Bray (2009), motivated by his conviction that the variation observable within the chemical data was not simply 'noise' but contained evidence of human behaviour. This resulted in the definition of what are now known as 'Copper Groups', and was combined with ubiquity mapping to study the movement of metal in the British and Irish Early Bronze Age. These ideas were subsequently developed over several years with the support of the Leverhulme Trust and the John Fell Fund, gradually being extended to deal with alloy data, and finally lead isotope data. In the meantime the opportunity arose to submit an application to the European Research Council for a large-scale project to study the circulation of metal in the Bronze Age across the whole of Eurasia, which was awarded and began in October 2015. The title was 'the *FLow of Ancient Metal across Eurasia*' (FLAME). Part of this project involved further developmental work on the interpretative systems, and the creation of a large open access GIS database of all of the known chemical and isotopic analyses of Bronze Age copper alloy metalwork from all of Eurasia north of the Himalaya. This database will also contain embedded tools which allow the implementation of the interpretative methods developed within the project.

This volume is not a summary of the outcomes of the FLAME project—that is still ongoing, and data continue to be entered into the GIS database. It is specifically intended to describe in some detail the methodologies proposed for interrogating these data, and to explain the philosophy underlying the whole project, since it is quite different from much of what has gone before. Several papers (listed at the end of this preface) have already been published which explain different aspects of the methodology, but inevitably they are rather brief because they are in the peer-reviewed journal literature. Moreover, our ideas have developed rapidly over the past five years, and sometimes the emphasis of how we see the system has changed. Here we present what we now see as an integrated system, and we also have the opportunity to explain the methodology and philosophy in much more depth than has been possible previously. We illustrate the methodology with some examples taken from recent work in Oxford.

FLAME and other relevant recent publications from Oxford

Liu, R., Pollard, A.M. and Rawson, J. (2018). Beyond ritual bronzes: multiple sources of radiogenic lead used across Chinese history. *Nature Scientific Reports* **8**, Article number: 11770.

Pollard, A.M., Rawson, J. and Liu, R. (2018). Some recently rediscovered analyses of Chinese bronzes from Oxford. *Archaeometry* **60** 118–127.

Y.-K. Hsu, J. Rawson, A.M. Pollard, Q. Ma, F. Luo, P.-H. Yao and C.-C. Shen (2018). Application of kernel density estimates to lead isotope compositions of bronzes from Ningxia, Northwest China. *Archaeometry* **60** 128–143.

Jin, Z.Y., Liu, R., Rawson, J. and Pollard, A.M. (2017). Revisiting lead isotope data in Shang and Western Zhou bronzes. *Antiquity* **91** 1574–1587.

Perucchetti, L. (2017). *Physical barriers, cultural connections: a reconsideration of the metal flow at the beginning of the metal age in the Alps*. Archaeopress, Oxford

Liu, R. (2017). *Capturing changes: applying the Oxford system to further understand the movement of metal in Shang China*. Unpublished DPhil Thesis, University of Oxford.

Liu, Ruiliang, A.M. Pollard, Jessica Rawson and Tang Xiaojia (2017). 共性、差异与解读: 运用牛津研究体系探究早商郑州与盘龙城之间的金属流通 (Revisiting the movement of metal between Zhengzhou and Panlongcheng in early Bronze Age China). *Jianghan Kaogu,* 111–121 (In Chinese).

Sabatini, B. (2017). *Chemical Composition, Thermodynamics, and Recycling: the Beginnings of Predictive Behavioral Modeling for Ancient Copper-based Systems*. Unpublished DPhil thesis, University of Oxford.

Pollard, A.M., Bray, P., Hommel, P., Hsu, Y.-K., Liu, R. and Rawson J. (2017). 牛津研究体系在中国古代青铜器研究中的应用 (Applying the Oxford System to further understand Bronzes in China). *Kaogu* (1), 95–106 (In Chinese).

Zhangsun, Y.Z.; Lui, R.L., Jin, Z.Y., Pollard, A.M., Lu, X., Bray, P.J, Fan, A. and Huang, F. (2017). Lead isotope analyses reveal the key role of Chang'an in the mirror production and distribution network during the Han dynasty. *Archaeometry* 59, 685–713.

Pollard, A.M., Bray, P., Hommel, P., Hsu, Y.-K., Liu, R. and Rawson, J. (2017). Bronze Age metal circulation in China. *Antiquity* 91, 674–687.

Pollard, A.M. (2017). Lead isotopes in archaeology. In Gilbert, A.S. (ed.) *Encyclopedia of Geoarchaeology*, Springer, pp. 469–473.

Pollard, A.M. (2016). The first hundred years of archaeometallurgical chemistry: Pownall (1775) to von Bibra (1869). *Historical Metallurgy* 49(1) for 2015 (published 2016) 37–49.

Hsu, Y.-K. (2016). *Dynamic flows of copper and copper alloys across the prehistoric Eurasian steppe from 2000 to 300 BCE*. Unpublished DPhil Thesis, University of Oxford.

Bray, P.J. (2016). The Saltonstall Early Bronze Age Axe. *Prehistoric Yorkshire* **53,** 99–103.

Bray, P.J. (2016). Metal, metalwork and specialisation: The chemical composition of British Bronze Age swords in context. In Koch, J. (ed.) *Celts from the West III*, Proceedings of the AEMA Conference, Cardiff 2014. Centre for Welsh and Celtic Studies, University of Aberystwyth.

Yiu-Kang Hsu, Peter J. Bray, Peter Hommel, A. Mark Pollard and Rawson, J. (2016). Tracing the flows of copper and copper alloys in the Early Iron Age societies of the eastern Eurasian steppe. *Antiquity* **90** 357–375.

Aurélie Cuénod, Peter Bray and A. Mark Pollard (2015). The 'tin problem' in the Near East – further insights from a study of chemical datasets on copper alloys from Iran and Mesopotamia. *Iran* **LIII** 29–48.

Pollard, A.M. and Bray, P.J. (2015). A new method for combining lead isotope and lead abundance data to characterise archaeological copper alloys. *Archaeometry* **57** 996–1008.

Perucchetti, L. (2015). *Physical barriers, cultural connections: a reconsideration of the metal flow at the beginning of the metal age in the Alps*. Unpublished DPhil Thesis, University of Oxford

Perucchetti, L., Bray, P., Dolfini, A. and Pollard, A.M. (2015). Physical barriers, cultural connections: prehistoric metallurgy in the Alpine region. *European Journal of Archaeology* **18** 599–632.

Ruiliang Liu, Peter Bray, A.M. Pollard and Peter Hommel (2015). Chemical analysis of ancient Chinese Bronzes: past, present and future. *Archaeological Research in Asia* **3** 1–8, July 2015 (doi:10.1016/j.ara.2015.04.002).

Pollard, A.M., Bray, P., Gosden, C., Wilson, A. and Hamerow, H. (2015). Characterising copper-based metals in Britain in the First Millennium AD: A preliminary quantification of metal flow and recycling. *Antiquity* **89** 697–713.

Bray, P., Cuénod, A., Gosden, C., Hommel, P., Liu, R. and Pollard, A.M. (2015). Form and flow: the 'karmic cycle' of copper. *Journal of Archaeological Science* **56** 202–209.

Pollard, A.M., Bray, P.J. and Gosden, C. (2014). Is there something missing in scientific provenance studies of prehistoric artefacts? *Antiquity* **88** 625–631.

Pollard, A.M. and Bray, P. (2014). Chemical and isotopic studies of ancient metals. In Roberts B.W. and Thornton C.P. (eds.) *A Global Perspective in Early Metallurgy*, Springer: New York, pp. 217–237.

Pollard, A.M. (2013). From bells to cannon – the beginnings of archaeological chemistry in the Eighteenth Century. *Oxford Journal of Archaeology* **32** 333–339.

Bray, P.J. and Pollard, A.M. (2012). A new interpretative approach to the chemistry of copper-alloy objects: source, recycling and technology. *Antiquity* 86 853–867.

Cuénod, A. (2013). *Rethinking the Bronze-Iron Transition in Iran: Copper and Iron Metallurgy before the Achaemenid Period*. Unpublished D.Phil. thesis, University of Oxford.

Bray, P.J. (2012) Before ^{29}Cu became *copper*: tracing the recognition and invention of metalleity in Britain and Ireland during the third millennium B.C. In M. Allen, J. Gardiner and A. Sheridan (eds.) *Is there a British Chalcolithic: people, place and polity in the later 3rd millennium*. The Prehistoric Society Research Paper 4: 56–70.

Bray, P.J. (2009). *Exploring the Social Basis of Technology: Re-analysing Regional Archaeometric Studies of the first Copper and Tin-Bronze use in Britain and Ireland*. Unpublished D.Phil. Thesis, University of Oxford.

In press

Pollard, A.M., Liu, R., Rawson, J. and Tang, X. (in press). From alloy composition to alloying practice: Chinese bronzes. *Archaeometry*.

In preparation

Liu, R. Pollard, A.M., Rawson, J., Tang, X. and Zhang, C. (in prep.). Panlongcheng, Zhengzhou and the Movement of Metal in Early Bronze Age China.

Howarth, P. (in prep.). *Copper circulation in Western Asia*. D.Phil. Thesis, University of Oxford.

Pollard, A.M., Liu, R. and Rawson, J. (in prep.). Every Cloud has a Silver Lining: using silver concentration to identify the number of sources of lead used in Shang Dynasty Bronzes.

Zhang, C., Rawson, J., Tang, X., Huan, L., Liu, R. and Pollard, M. (in prep.). Reappraisal of the origins of China's Bronze Age.

Acknowledgements

The work presented here has been under development for more than 10 years, and has received support from various sources, including a Hastings Senior Scholarship given by The Queen's College Oxford to Peter Bray, three awards from the Oxford University Press John Fell Fund (1 November 2008, 1 December 2009 and 1 April 2011), and the Leverhulme Trust (Grant F/08 622/D: Pollard, Gosden and Northover 'Chemical structure and human behaviour: a new model for prehistoric metallurgy', 1/01/11). The main source of support has been European Research Council Advanced Grant 1300505 FLAME (**FLow of Ancient Metal across Eurasia**), from 1/10/2015–30/9/2020.

We are grateful for the support given to the project by the FLAME project partners:
Prof. Christopher Bronk Ramsey, RLAHA, Oxford
Prof. Chen Jianli, University of Beijing
Prof. Chen Kunlong, University of Science and Technology, Beijing
Prof. Sir Barry Cunliffe, Institute of Archaeology, Oxford
Prof. Chris Gosden, Institute of Archaeology, Oxford
Prof. Dame Jessica Rawson, Institute of Archaeology, Oxford
Prof. Mei Jianjun, Needham Research Institute, Cambridge
Prof. Rüdiger Krause, Goethe University, Frankfurt
Prof. Viktor Trifonov, Institute for the History of Material Culture, St Petersburg
Dr. Natalia Shishlina, State Historical Museum, Moscow
Prof. Tsagaan Turbat, Mongolian Academy of Sciences, Ulaanbaatar
Prof. Jin Zhengyao, Archaeometry Laboratory, University of Science and Technology of China, Hefei

We are also grateful to Steve Stead, for his support and advice in constructing the GIS-database. Finally, we owe a huge debt to those who have worked and published in this field before us, and particularly to Evgenij Chernykh: aspects of our model, presented here, parallel much of his work published in the 1960s-1990s, although we have come to it by a very different route. We also acknowledge all those who have produced and published chemical and isotopic data over the last 200 years, without which such a project would be impossible.

Table of Contents

Chapter 1

Previous Approaches to the Chemistry and Provenance of Archaeological Copper Alloys

For more than two million years, hominins have exploited some of the rich mineral resources of the earth's crust. These have provided useful raw materials, such as particular stones for tools and weapons, pigments for painting and body art, brightly coloured minerals and semi-precious stones for body ornamentation, clays for figurines and ceramics, and, more recently, ores from which to extract metals. Although rich, such mineral resources are usually concentrated in particular regions, which tend to be located in mountainous or desert areas, and are hence difficult to reach. In spite of this, humans in the past must have specifically sought out such deposits, and probably transported some of their products over long distances, giving rise to systems of trade and exchange in particular precious commodities. Some such systems can be traced back over millennia, particularly those involving lithics with desirable qualities, such as being decorative or having the capacity to be worked into stone tools. It seems certain that metals, once they began to circulate within society, would have fallen into this class of desirable material. As people began to understand their unique properties, metals transformed human relationships with the material world by providing efficient tools and weapons, but also by acting as symbols of wealth and power.

Unsurprisingly, the origins of metallurgy have been the focus of intense academic speculation for several centuries. Although it was perhaps preceded by the smelting of lead (Krysko 1980; Gale and Stos-Gale 1981), locating the time and place of the first exploitation of copper and its alloys has been a major focus of academic endeavour. It now seems reasonably certain that the first smelted copper in Eurasia is to be found in or near Anatolia, perhaps in the seventh millennium BCE, and spreading to South-East Europe and Western Asia by the fifth millennium BCE (Roberts *et al*. 2009; Kienlin 2013). The knowledge of copper smelting subsequently spread to all of Eurasia by the second millennium BCE, via a process seen by most as a form of direct transmission. An independent origin is presumed for the smelting of copper in the New World, and has also been suggested for some parts of Eurasia, particularly China, but there is little evidence for this. It is interesting to note that the smelting of copper ore in Anatolia and Western Asia appears to have been preceded by a long period of the use of brightly coloured copper minerals (e.g., malachite, azurite) for personal ornamentation (and occasionally as part of burial rituals), perhaps for several millennia (e.g., Solecki 1969). The desire for such colourful minerals would have led people to search for ore deposits in the mountains, in turn leading to the discovery of occasional rare finds of native copper, and ultimately to the smelting of the ores. Perhaps this was initially a result of some accidental heating of a mineral such as malachite, although the required temperature is higher than might easily be conceived of as 'accidental', and it may therefore have been the consequence of a more systematic experimental inquisitiveness. This sequence of malachite use might suggest an answer to the oft-posed question of why the remarkable Neolithic civilizations of East Asia came so late to the use of copper. The common consensus is that copper and copper smelting did not arrive in central China until the beginning of the second millennium BCE, and was not the result of independent invention. Since the Chinese Neolithic cultures valued jade above all other natural materials for symbols of ritual and power, the same trajectory of malachite - native copper - smelted copper is unlikely to have occurred, and therefore copper smelting was introduced into China from the west. What they did with it subsequently was, of course, remarkable and entirely Chinese. It could even be argued that the whole system of Chinese metallurgy, dominated by casting techniques which depend on metal fluidity, was a consequence of this late arrival, and was based on the notion of metal *as metal* rather than emerging from earlier techniques adapted from the physical working of stone.

As stated, the FLAME project is explicitly *not* concerned with the origins of smelted copper. The primary concern is to better understand the use and circulation of metal, and the development of metallurgical traditions, as these new materials became an intrinsic part of the fabric of prehistoric societies across Eurasia. Broadly speaking, this focusses primarily on the third and second millennia BCE, but since the chronology of the Bronze Age varies markedly across Eurasia, these do not represent strict chronological boundaries for the project. It does mean, however, that we *are* interested in the rise and spread of copper alloying technologies across Eurasia, from arsenical coppers/bronzes, to tin bronze and leaded tin bronze, since their development and transmission falls broadly within this period.

Chemical analysis of archaeological copper alloys

Chemical analysis as we know it today was developed in Europe towards the end of the eighteenth century, but that does not mean that the composition of metal objects was unknown and unknowable before then. The art of assaying the chemical composition of precious metal objects has probably been practised for almost as long as the use of the metals themselves. Greenaway (1962) emphasises that assaying—testing by fire—is the oldest quantitative chemical technique. From written sources it can be traced back to the early 2nd millennium BCE in Mesopotamia, where cuneiform tablets describe in some detail the quantitative assay of gold and silver (Levey 1959), but these techniques were probably old even then. Surviving European medieval texts give us increasingly clear descriptions of the process of assaying, both of metal objects and, in the case of base metals, of the ores from which they come. Book III of Theophilus' *On Divers Arts* (c. 1110–1140 CE; Hawthorne and Smith 1963) describes how to recover gold from scrap gilded metal, and also how to part gold from silver. The *Probierbüchlein* ("*The little book on assaying*"), written by an unknown German goldsmith or assayer around 1520 CE, is the first western book to give a clear description of assaying (Sisco and Smith 1949). Lazarus Ercker's *Beschreibung allerfürnemisten mineralischen Ertzt und Berckwercksarten* ("Description of all forms of minerals and calcareous species"), originally published in Prague in 1574, rapidly became disseminated throughout Europe, and was reprinted in 1580, 1598 and 1629. It was translated as "*Lazarus Ercker's Treatise on ores and assaying*" by Sisco and Smith (1951).

The art of using the touchstone to assess the fineness of precious metals, especially gold, is probably equally as old as assaying (Oddy 1986). A touchstone is a flat piece of dark fine grained stone upon which the metal is rubbed, with the colour of the streak indicating the purity of the gold. From the calculations made by Oddy based on Theophrastus' book "*On Stones*" (*c*. 315 BCE; Caley and Richards 1956, 54), the touchstone could detect one part in 144 of impurity in the gold, which is a sensitivity not surpassed until modern times. Oddy (1986, 164) says that the first certain reference to the touchstone is in the 6th century BCE, but the fact that it was so well known to Theophrastus, and that it is also referred to in Sanskrit texts contemporary to Theophrastus, suggests that it is much older than this.

The origins of analytical chemistry as we understand it today are to be found in 18th century Europe, when 'trial by fire' gave way to 'the humid method', or precipitating known compounds out of solutions, ultimately leading to quantitative gravimetric analysis. Through the technical development of the analytical balance, and increasingly systematic study of aqueous chemical reactions and precipitations by Robert Boyle, Étienne François Geoffroy, and others, Torbern Bergman at the

University of Uppsala, Sweden, published in 1777 a protocol for the aqueous gravimetric analysis of gemstones. This was followed by more detailed protocols from Nicolas Louis Vauquelin in Paris (1799) and Martin Heinrich Klaproth in Berlin (1792/3). The methods of these three pioneer analytical chemists have been described and compared by Oldroyd (1973).

One of the earliest applications of the new method of gravimetric analysis was to archaeological material, probably as part of a broader general interest in the contents of the 'cabinets of curiosities' of the time. Thus, Dizé published the analysis of the tin content of some copper alloy coins and other objects in 1790, and Klaproth analysed a wider selection of coins, as well as glass, in 1792/3 (Pollard 2013). In these early stages, the purpose was simply to identify the metals used in these alloys. This was in part prompted by the mid-18th century debate in France over whether bronze was an alloy of copper with iron, or copper with tin (Pollard 2013). These first chemical analyses clearly showed that bronze was an alloy of copper with tin. The discussion then moved on towards understanding changing patterns of alloy use over time and space, most clearly visible in the work of Göbel (1842), and ultimately crystallising as the concept of 'provenance'.

The provenance hypothesis

The observation that raw materials (stone, obsidian, clay, metals) are likely to have been obtained from specific and restricted geographical source areas, and potentially transported over long distances, has given rise to one of the major continuing themes in scientific archaeology, that of *provenance*. Essentially, provenance studies are based on the assumption that some characteristics of the source of the raw material are carried over into the finished object, and that they are sufficiently diagnostic to allow differentiation between geographically distinct sources (Wilson and Pollard 2001). These characteristics can be trace element patterns, rare earth profiles, or isotopic ratios, and are often referred to as a chemical *fingerprint*. The idea that patterns of chemical composition in archaeological artefacts could be used to attribute source had become well established by the mid-19th century in Europe (Pollard *et al.* 2014). Göbel (1842) noted in his monograph (entitled, significantly, "*About the influence of chemistry on the determination of the peoples of the past, or results of the chemical analysis of metallic antiquities...*") that some Roman copper alloys contained zinc, whereas Greek alloys contained only tin and some lead. Even more significantly, he used the available analyses of Roman coins to divide them up into four groups (Cu + Sn + Pb, Cu + Zn, Cu + Zn + Sn, and Cu + Sn + Zn + Pb), and, with the exception of the first group, noted that the amount of zinc decreased over time. More than a hundred years later, Caley (1964a) also noticed that the zinc content of Roman brass coins declined from the late 1st century BCE to the early third century CE. He suggested that Roman brass production started in the late 1st century BCE but stopped shortly after, and that brass coins of the late 1st century CE onwards were made from recycled brass. This 'zinc decline' was thus thought to be a consequence of the volatility of zinc, which meant that the zinc content of the brass declined after each re-melting (Caley 1964a, 99).

Göbel (1842) thought that the chemical differences he observed could be used to date metal objects from regions outside Greece and Rome. He thus compared the analyses of metals found in graves in the Baltic States with Roman coinage of particular Emperors, in an attempt not only to elucidate the origin of the metal, but also to give a date to the as then undated metal objects from northern Europe. Altogether, Göbel published the analyses of 119 archaeological artefacts, which was the first large-scale attempt to study the chemical composition of ancient metal objects. The first very large compilation of chemical analyses on archaeological metals was published by von Bibra (1869), containing approximately 1250 analyses, of which 600 were his own, and the other 650 were taken from the earlier work of at least 90 other analysts, including Göbel.

The theory of provenancing archaeological materials by chemical means was first fully articulated by Damour (1865, 313):

> "*When one discovers, in fact, either buried under the ground, either in the caverns, or among the remains of ancient monuments, an object on which the hand of man has marked his work, and whose matter is of distant origin or foreign to the country, it is inferred that there has been transport of the object itself, or at least of the matter of which it is formed. Hence arises inductions on the relations which may have existed between different peoples, on their migrations, their industry, etc.*"

Until recently, this statement has essentially remained the main theoretical underpinning for much of the work involving the determination of provenance of archaeological materials using chemical analysis, including metals, ceramics, glass and lithics (Wilson and Pollard 2001). It is has been elaborated upon for specific materials (e.g., for metals, Tylecote 1970; Pernicka 1986, 1999; Budd *et al.* 1996), but it is only recently that theoretical considerations such as the influence of the time taken for objects to move between source and deposition site have been explored in more detail (Pollard *et al.* 2014).

Major European programmes of chemical analysis

Although gravimetric analysis continued to be the preferred method of many distinguished metallurgists until at least the 1960s (e.g., Caley 1964b), a major step-change in the analysis of archaeological metals took place in Europe in the early 20th century, and accelerated considerably after the Second World War. This was brought about primarily because of the increased availability of instrumental means of chemical analysis, initially using optical emission spectroscopy (OES: Smith 1933; see Pollard *et al.* 2017, 34–35), which meant that many more samples could be analysed. It also increased the range of trace elements that could be quantified. As a consequence, several large European projects were initiated with an explicit focus on the provenance of the copper used to make Bronze Age archaeological metal artefacts. The earliest in the UK followed the establishment of the Ancient Mining and Metallurgy Committee of the Royal Anthropological Institute in 1945 (Anon. 1946). This Committee was an interdisciplinary panel consisting of some of the leading archaeologists of the day, including V. Gordon Childe (1892–1957), Oliver Davies (1905–1986), Christopher Hawkes (1905–1992) and Stuart Piggott (1910–1996), and scientists E. Voce (1902–1960), Cecil Henry Desch (1874–1958), Harold J. Plenderleith (1898–1997), Cyril E. N. Bromehead (1885–1952) and Herbert Henery Coghlan (1896–1981). The stated aim (Coghlan *et al.* 1949, 6) was summarized as:

> "*The main question now before this committee is whether there are any means of recognizing the locality from which the metal in a given copper object was obtained. The information already available on the subject has been conveniently assembled, with full references, by J. R. Partington* (Origins and Development of Applied Chemistry, 1935)."

The Ancient Mining and Metallurgy Committee published a series of more than 20 papers from 1948 to about 1957, but the deaths of several of the leading members of the Committee in the 1950s seems to have brought about the demise of the initiative (although some, particularly Coghlan, continued to publish through to the 1980s).

Otto and Witter

A little earlier than this, in the 1930s, several groups of German-speaking researchers began large scale programs of chemical analyses on ancient metal artefacts. The first significant compilation of data was that of Helmut Otto (1910–1998) and Wilhelm Witter (1866–1949), from the University of Halle, whose results were summarized in 1952 as "*Handbuch der ältesten vorgeschichtlichen Metallurgie in Mitteleuropa*". Their aim was to understand prehistoric metallurgy in Europe using chemical analysis, in contrast to the previous typological approach to metal artefacts. They decided to focus on only one period to increase the consistency of their work, namely the very beginning of the Bronze Age. They organized a program of research to initially analyse a substantial number of artefacts (1374 artefacts, of which ca. 1100 were analysed by Otto, and the remainder by J. Winkler, W. Noddack and W. Kroll) from all across Europe, from Ireland and Denmark to Italy and from Spain to Romania (Otto and Witter 1952, 1–21).

They developed their own optical emission spectrometry (OES) methodology that allowed them to obtain quantitative analyses: a sample of approximately 0.2 g of unaltered metal from the artefact was melted to form two electrodes, and then a high voltage was applied between them, causing the emission of light whose wavelength was dependent on the chemical elements in the sample. The intensity of the light emitted gave information about the quantity of each element present in the sample. Moreover, they also emphasized the importance of metallography to study the manufacturing process of the artefacts, giving an input into a programme of creating experimental artefacts.

From their analyses of more than 1300 Early Bronze Age artefacts they created six groups of metal according to the artefacts' compositions: "pure copper (Reinkupfer)", "raw copper (rohkupfer)" (which indicates "copper with small traces of other elements"), "arsenical copper alloy (arsen-kupferlegierung"), "Fahlerzmetalle" [1], divided into "Fahlerz with a high percentage of silver" and "Fahlerz with a low percentage of silver", "other kinds of metal (sonstige metalle mit Ni, As und Ag)", and "copper tin alloy (zinn-kupferlegierung)". These groups were further divided according to the percentage of presence of tin, lead, silver, gold, nickel, cobalt, arsenic, bismuth, zinc (Otto and Witter 1952, tab. I–VI). The creation of these groups was based only on the authors' observations, without any statistical treatment of the data. Although their ability in the visual identification of these groups was subsequently recognized by later authors (e.g., Ottaway 1982, 94–95), such a subjective technique was not thought to be acceptable for scientific research and was soon replaced by statistical analysis (see below). Finally, these authors also investigated chronological and geographical patterns linked to their groups to hypothesize the provenance of the material and to identify trade routes (Otto and Witter 1952, 60–82). As regards chronology, they accepted Childe's theories of a linear technological evolution in metallurgy, from copper, to arsenical copper, Fahlerz, and tin bronze (Childe 1944) and dated the objects according to their chemistry (Otto and Witter 1952, 5). With respect to provenance, they

[1] Fahlerz was defined by them as "copper with a higher percentage of trace elements than raw copper."

claimed that almost all of the German artefacts came from copper ores in Saxony, but this conclusion has subsequently been challenged (Pernicka 2011, 28).

Pittioni and the Vienna group

Another important contemporary group of researchers was established in Vienna in the 1930s, led by Richard Pittioni (1906–1985). This group was also interested in the provenance of ancient metal artefacts through compositional analysis, and focused their research on the Alpine and Balkans regions (Preuschen and Pittioni 1937). They brought a strong metallurgical background to the study, since they were both mining engineers (Pernicka 2011, 28). They argued against the methodology of Otto and Witter, pointing out that only with a significant amount of data of the same period and of specific forms can a consistent hypothesis be formulated about the provenance of the ore used for a group of artefacts (Pittioni 1957, 3). Furthermore, they declared explicitly that it is quite impossible to understand the provenance of a single object because metal ores are too heterogeneous, and the composition of the copper is not the same at all depths within the ore deposit. Consequently, two objects with different compositions could derive from two different points in a single ore source. The heterogeneous nature of metal ores also implies that the presence of a single specific chemical element in an artefact is never crucial in characterizing metal groups (Pittioni 1957, 4). Useful information can only be derived by the combination of the presence of some specific elements: antimony, arsenic, lead, nickel, silver, bismuth and tin (Pittioni 1957, 7). Iron was considered not to be diagnostic, as it is universally present in copper ores; aluminium, calcium, magnesium and silicon were also ruled out as non-specific (Pittioni 1957, 7). A crucial consequence of the Viennese group's idea of the combination of presence/absence of elements as being the main priority in data collection was their decision to undertake only semi-quantitative analysis, without giving numerical values for their concentrations (**Table 1**).

This approach was heavily questioned at the time (see discussion following Pittioni 1960), and, unfortunately, it also means that their data are not useful for modern research using a statistical perspective. This was pointed out by Ottaway (1982, 175–176), who tried to convert their data expressed as symbols into numerical data. As discussed in Chapter 3, although her efforts were successful with the more limited set of symbols used by Otto and Witter and the Stuttgart group (see below), the data from Pittioni's work were impossible to convert and, hence, could not be included in her database. For this reason, in the end, more than 6000 valuable analyses are now completely unusable. Despite this considerable drawback, the group from Vienna deserve credit for pointing out the need for a large number of contemporary objects in order to theorize provenance, and also for recognizing the importance of a multidisciplinary approach in archaeometallurgical research, including the contribution of geology. Indeed, the Vienna group analysed not only finished artefacts, but also slags and metal ores. Consequently, they initiated geological research in the alpine region to identify the possible ore sources and to look for sites that could provide evidence for ancient smelting processes (Pittioni 1957, 7–16).

Nr	Objects	Cu	Sn	Ag	Al	As	Ca	Fe	Mg	Mn	Ni	Pb	Sb	Zn	Bi	Au	V
Barreringen A																	
1	Metallkern 1	HM	Sp	++	-	++	-	++		Sp	-	Sp	+	-	+	-	-
2	Metallkern 2	HM	Sp	++	-	++	-	++		Sp	-	Sp	+	-	+	-	-
3	Oberfläche 1	HM	Sp	++	-	++	-	++		Sp	-	Sp	+	-	+	-	-
4	Oberfläche 2	HM	Sp	++	-	++	-	+		Sp	-	Sp	+	-	+	-	-
Barreringen B																	
5	Metallkern 1	HM	Sp	++	-	++	-	++	-	-	Sp	Sp	+	-	+	-	-
6	Metallkern 2	HM	Sp	++	-	++	-	++	-	-	Sp	Sp	+	-	+	-	-
7	Oberfläche 1	HM	Sp	++	-	++	-	+	-	-	Sp	Sp	+	-	+	-	-
8	Oberfläche 2	HM	Sp	++	-	++	-	+	-	-	Sp	Sp	+	-	+	-	Sp
Barreringen B																	
9	Metallkern 1	HM	+	++	-	++	-	++	-	-	Sp	Sp	+	-	+	-	-
10	Metallkern 2	HM	+	++	-	++	-	+	-	-	Sp	Sp	+	-	+	-	Sp
11	Oberfläche 1	HM	+	++	-	++	-	+	-	-	Sp	Sp	+	-	+	-	Sp
12	Oberfläche 2	HM	+	++	-	++	-	+	-	-	Sp	Sp	+	-	+	-	Sp

Table 1:
Example of the data published by Pittioni (1957, Tab. 1).

Stuttgart and the SAM project

Around the middle of the 20th century a project was started in Stuttgart by a group of researchers, the most eminent of whom were Siegfried Junghans (1915-1999), Edward Sangmeister (1916-2016) and Manfred Schröder (1926-2009). Their work is usually referred to by later authors, and also here, as "the SAM project", in abbreviation of the full title, "*Studien zu den Anfangen der Metallurgie*". The explicit aim of their research was to study the origin and spread of copper and bronze in Europe by scientifically examining the material itself, by means of optical emission spectroscopy and statistical analysis (Junghans *et al.* 1968, 6). Their field of research was broader than that of their predecessors: for the first time the entire European continent was taken into consideration. They created European distribution maps of objects with common chemical compositions. The extent of the territory of interest and the necessity to have a statistically valid sample caused the authors to make as many analyses as possible, ultimately producing more than 22,000. They published the first 1000 results in 1960, added 9,000 more in 1968 and finally reached 22,000 in 1974. In this last publication, they also reconsidered their previous analyses, republishing some of them because in some cases arsenic and antimony had been underestimated by their instrument (Junghans *et al.* 1974). Fortunately, following in the footsteps of Otto and Witter, they produced fully quantitative analyses.

They focused on eleven elements: tin, lead, arsenic, antimony, silver, nickel, bismuth, gold, zinc, cobalt, and iron. The quantity of copper was not determined directly. The authors recognised the most distinct boundaries between different types of copper in the percentages of bismuth, antimony, silver, nickel and arsenic (Junghans *et al.* 1960, 57). From this, Hans Klein, the statistician of the group, used these elements to develop his statistical frequency analysis. He tried to create groups in which the frequency distribution of each element could be represented as a Gaussian (normal) curve. As a result, Klein defined 12 groups of metal, as shown in **Figure 1** (A, B1, B2, C1, C2, C3, E00, E01, E10, E11, F1, F2). A group was considered to be "secure" only when it reached a threshold that did not change even with an increasing number of samples (Junghans *et al.* 1960, 58). However, they noted that, due to the statistical method applied, a few samples very close to the boundaries of the groups could be assigned to either group (Junghans *et al.* 1960, 58).

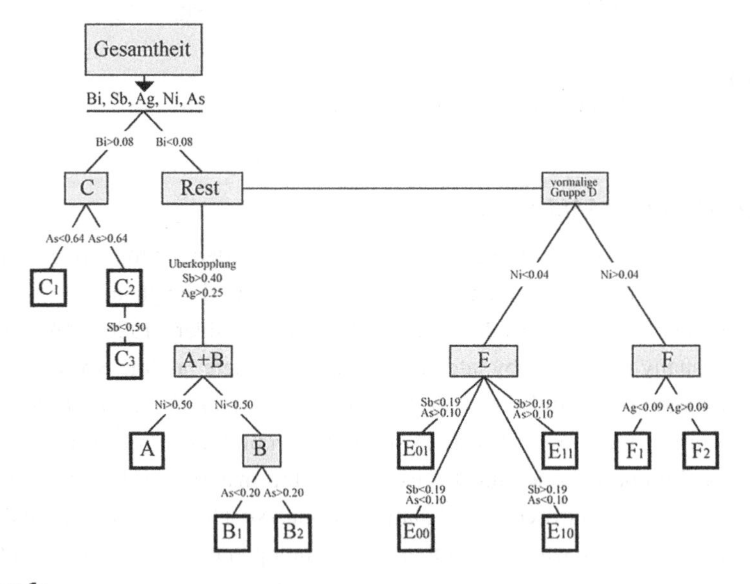

Figure 1:
Groups of metal artefacts according to their composition, re-drawn after Junghans ***et al.*** (1960, Tabelle 1, p. 210)

When the Stuttgart group published their second set of analyses in 1968, they decided to perform their statistical analysis by plotting each element against the other elements, which they called "two dimensional analysis" (Junghans *et al.* 1968, 13). With this technique, they created a first subdivision of artefacts into groups, which was then refined by considering the percentages of individual elements, in particular Ag, As, Bi, Ni and Sb. As a result, instead of the original 12 groups, 29 groups were created, as shown in **Figure 2**. However, most of the groups already identified in the 1960s publication were confirmed, and it was felt that the new results only helped to further refine these groups (Junghans *et al.* 1968, 15).

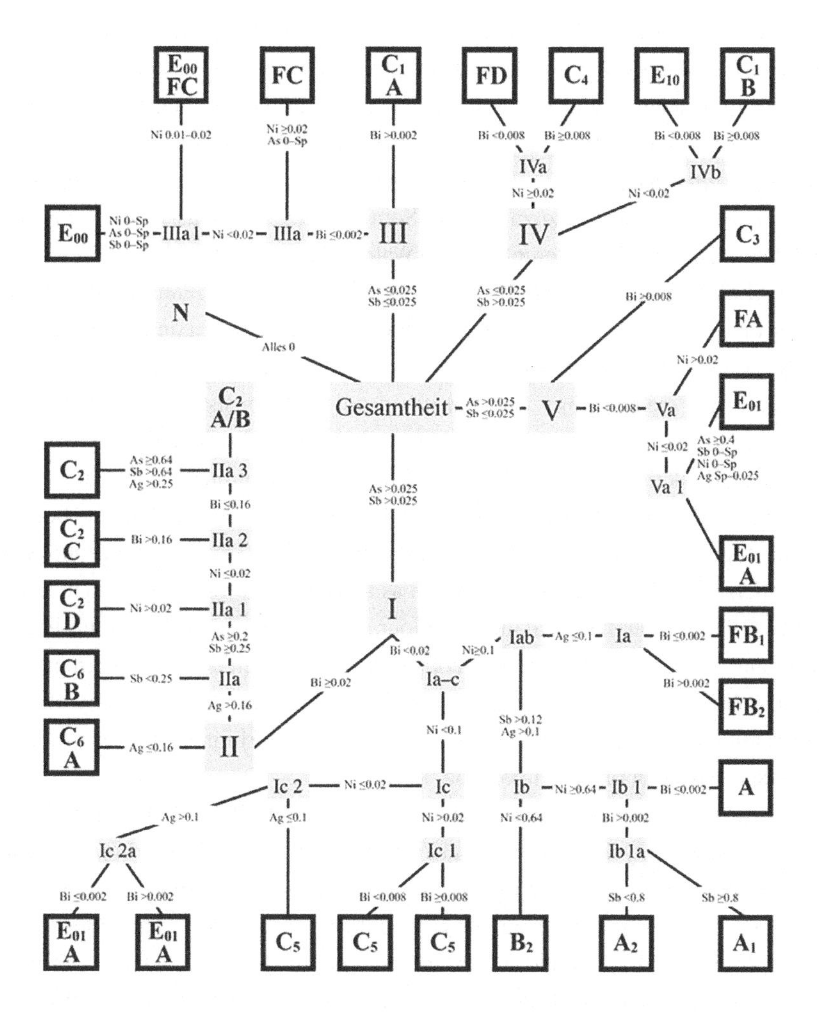

Figure 2:
SAM 'decision tree' redrawn after Junghans ***et al.*** (1968) (SAM 2.2, Tabellen und Diagramme, Diagramm 1).

Nevertheless, as stated by Muhly (1993),

> "*the reaction to the SAM Project was uniformly negative, ... archaeologists were sceptical of the SAM metal groupings chiefly because the different classes of metal were presented within the context of an outmoded diffusionist archaeology that was (unfairly) used to discredit the entire project*."

In particular, the statistical methodology of the Stuttgart group was heavily criticised, most specifically by Waterbolk and Butler (1965), who argued against both the methodology and the presentation of their results. They strongly criticized the lack of consideration of the archaeological background in their methodology. Waterbolk and Butler (1965, 230) stated that the Stuttgart team had "*thrown the analyses all into one pot, with the hope that mathematical means will bring them out of the pot again in a logical order*." Some artefacts were assigned to groups with characteristics of a different period (Waterbolk and Butler 1965, 232). Other groups of objects were considered as belonging to several different groups, even though they were from a single homogenous archaeological context and had a mostly homogenous composition (Waterbolk and Butler 1965, 237). Another criticism of the methodology was that it had not been consistently applied to all the defined groups: in fact only group A had all elements distributed in a Gaussian curve (Waterbolk and Butler 1965, 231). Moreover, they pointed out that in certain cases the significant information was not the percentage of an element, but its ratio to other elements (Waterbolk and Butler 1965, 238). The Stuttgart group was also criticized for the adoption of bismuth as a discriminating element. Slater and Charles (1970) pointed out that the behaviour of bismuth, with its low solubility in copper and lower melting point than copper, leads to a high degree of segregation during solidification, which causes unreliable analyses. Consequently, these authors felt that this element should not be chosen to discriminate between groups of artefacts based on their composition.

In terms of data presentation, Waterbolk and Butler (1965, 233) also pointed out that if new analysis were undertaken it would be difficult to compare these results with the proposed groups, so that judging the probability of the artefact belonging to one group or another would be impossible. The alternative methodology proposed by these two authors was to start with artefacts from a homogenous archaeological context, such as a hoard, and represent graphically the distributions of the elements considered to be important (e.g., **Figure 3**). These criticisms of the Stuttgart group's statistical methodology have been generally accepted, but, as pointed out by Ottaway (1982, 97), the methodology proposed by Waterbolk and Butler is hardly usable with large complex datasets. However, the acceptance of such critiques does not undermine the fundamental importance of the Stuttgart group's work. A critical point was the decision to adopt the analytical methodology proposed by Otto and Witter, rejecting Pittioni's use of qualitative data. For the European Bronze Age, these data remain the largest and most comprehensive dataset available, which subsequent workers have attempted to re-interpret and evaluate (e.g., Krause and Pernicka 1996: see below). The full database was published electronically by Krause (2003), and this forms the core of the FLAME data for Europe.

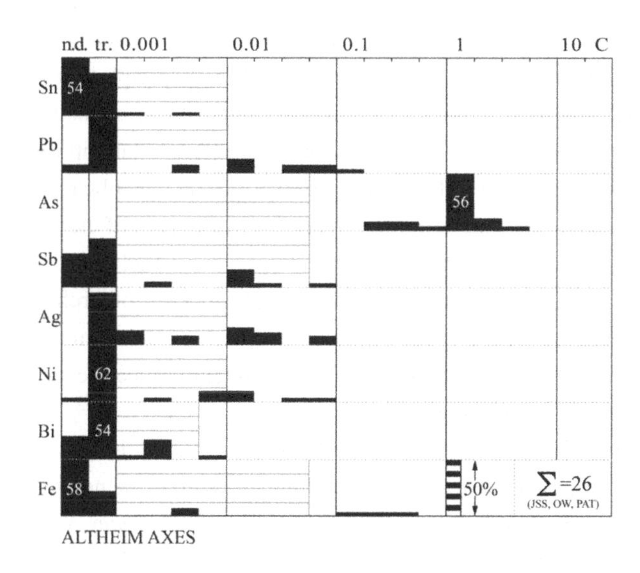

Figure 3:
The composition of 26 Bronze Age axes from Altheim as presented by Waterbolk and Butler (1965, Graph 3).

Evgenij Chernykh in Moscow

Until the 1960s, the history of archaeometallurgical research in Russia and the Soviet Union followed much the same pattern as was seen across the rest of Europe. Indeed, many of the early analyses of archaeological artefacts from Eurasia were carried out by the very same researchers who were active in Europe (e.g., Struve 1866). As elsewhere, the growth of quantitative spectrometric methods during the early part of the 20th century—which provided complete, rapid analysis at low cost, on small samples with little or no pre-treatment—led to an increase in the pace of archaeological research. By the end of the 1950s dedicated laboratories had been founded in Baku (Azerbaijan), Tbilisi (Georgia), St Petersburg and Moscow. Each of these produced substantial quantities of data and led to important regional and macro-regional studies. The most significant change, however, came with the foundation of a second analytical laboratory in Moscow within the Institute of Archaeology of the Soviet Academy of Sciences under the direction of Boris Kolchin, and associated from its inception with the work of Evgenij Chernykh. The subsequent work of this laboratory, which for prehistoric archaeology was driven forward by Chernykh, is known to most Western researchers through a single book—*Ancient Metallurgy in the USSR*—published in English in 1992, only a few months after the structure of the Soviet Union itself had collapsed. Intended as an accessible English summary of three decades of intensive analytical and archaeological work, this unintentional epitaph to a fallen empire (described as such by Chernykh 2017) was simultaneously praised for its clarity and scope as the first extended presentation of Soviet scholarship on this subject and harshly (though quite unfairly) criticized for its non-publication of data, its lack of analytical detail, and its perceived origins in the work of the Stuttgart group (see, for example, the review by Muhly (1993)). Nevertheless, it has remained the primary citation for most English-speaking students of archaeometallurgy in northern Eurasia. Certainly, in formulating our own ideas, this text remained almost our only window onto Russian archaeometallurgical research until the inception of the FLAME project.

Most English language descriptions of his work, including those of Chernykh himself, have focussed only on his archaeological results—the networks of central production and peripheral exchange, or *Metallurgical Provinces*, which shape the structure of his interpretations (see Chernykh 1992, 7–10; Kohl 2007; Chernykh 2008). It is worth, however, highlighting some of the fundamental justifications of his approach, which, far from being a mere Russian reflection of the SAM project, is one which emerged from a comprehensive critique of the existing methodologies in European archaeometallurgy. The complexity and subtlety of his approach are only apparent from a detailed study of his original Russian language publications.

Although Chernykh placed the question of provenance at the centre of his archaeological interest, he did not assume that it was straightforward, either from a geological or a metallurgical perspective. In the initial presentation of his methodology, Chernykh (1966, 13–17) reviews the problems of differentiating ore sources, using Russian data as an example. He concludes that while a general insufficiency of data on ore sources, from both archaeological and geochemical perspectives, usually reduces the theoretical scope of provenance studies to the regional level, we should not assume that the reliable differentiation of specific ore sources is impossible—a position he later developed in his research on mining in the Urals and Bulgaria (Chernykh 1970; 1978; 2004; 2002-2007). However, he was equally interested in understanding chemical change between ores and objects as a result of human action. Rejecting Thompson's (1958) negative assessment of the possibility of relating objects and ores, he was instead inclined to follow Biek (1957) in recognizing that arguments about relationships between archaeological ores and artefacts are typically sustainable only at the general level and not in individual cases. He nevertheless made a thorough review of the metallurgical literature, including experimental studies of the preferential movement of elements between metal, slag and vapour (Okunev 1960), to establish a reasoned baseline for the choice of key elements in his analysis (Chernkyh 1966, 18–21): Sn, Pb, As, Sb, Bi, Ag, Au, Co, Ni and Zn. Using both geochemical arguments and explorations of elemental distributions from a large number of analyses he established a coherent approach to the definition of the boundary between artificial and natural alloys (e.g., at around 1% for tin). Throughout this discussion, he also provided basic notes on several key problems relating to the recycling of metal, including the differentiation of primary alloys and alloyed metal resulting from the re-melting of scrap bronze with clean metal, and the impact of oxidative loss of particular elements on the overall composition of the metallurgical group. He concludes:

> "*One of the most complex and difficult tasks is the identification of secondary, mixed metal [within the system]. Such [metal] derives from the re-melting of broken artefacts, made from metal smelted from different ores... and containing a complex array of elements... derived from its [original components]. Evidently, in some archaeological cultures it is possible to identify such mixed groups... [but] the methodology by which to differentiate this metal is not entirely clear*" (Chernykh 1966, 20–21).

What is equally important, but often missed in English-language discussions of his methodology, is his integration of archaeological and chemical information within a standardized statistical approach, described more fully in his second thesis (Chernykh 1970). The first step is to define and codify various groups:

- Chemical—based on a characteristic suite of natural components—defined through a combination of visual and chemical analyses— and deemed distinctive of a particular region, mineralogical formation, or mine,
- Metallurgical—based on characteristic alloying components (e.g., Sn>1%) and independent of chemical groups,
- Typological—based on various characteristics and proportions within broad functional-stylistic groupings.

Up until 1989, the chemical analyses were made by OES using 5-10 mg of sample, and recording originally on a photographic plate, but employing two or three exposures at different excitation currents to ensure adequate recording of the different elements. He states in Chernykh and Lun'kov (2009) that their laboratory produced around 40,000 analyses by this method, of which so far only about half have been fully published. With the collapse of government funding at the end of the 1980s, the primary focus of the laboratory shifted towards the publication of these data and their synthesis with the growing body of radiocarbon results, a task which is still ongoing. After 2007, analytical work at his laboratory resumed using a desk mounted pXRF instrument, primarily on sampled material.

His approach to these chemical data was to assign the compositions to a limited number of metallurgical (alloy recipes) and chemical groups (trace elements) on the basis of visual examination of the data, which are specific to each area of study. An example of his classification of Eneolithic Bulgarian copper objects with Sn<1% is shown in **Table 2**, where the data are divided into six groups, labelled 1-6 (Chernykh 1978). In other areas he named the groups on the basis of their presumed association with sources or source regions (e.g., Chernykh 1970). His analysis of the data, however, then explicitly combines typological and chemical analysis. He carries out two sets of correlations between assemblages, one for typology (R) and one for chemistry (S). For typology, he defines a set of typological categories, and allocates every object in the assemblage to one of these categories, producing a numerical summary of how many objects belong to each category for each assemblage. He then compares assemblages on a pair-wise basis using the formula:

$$R_{AB} = \sum_{j=1}^{n} \sqrt{\frac{k_j m_j}{KM}}$$

where R is the correlation between assemblages A and B, k_j and m_j are the numbers of objects in groups A and B respectively classified into group j, and K and M are the total number of objects in assemblages A and B. He performs exactly the same correlation (S) for the objects classified into chemical groups. He then compares both correlations between all of his cultural assemblages in a correlation table, which is presented graphically as a series of columns, showing the relationship between one cultural assemblage and all the other assemblages in his analysis (**Figure 4**). This allows an evaluation of the relationship between cultural assemblages for both typological and chemical data. This approach forms the basis of his derivation of metallurgical provinces.

	I	II	III	IV	V	VI
Sn	0–0.005	0–0.005	0–0.01	0–0.01	0–0.01	0–0.01
Pb	0–0.01	0–0.01	0.001–0.3	0.001–0.1	0.01–1	0.001–0.01
Zn	0–(0.008)	0–(0.008)	0–(0.008)	0–(0.008)	0–(0.008)	0–(0.008)
Bi	0–0.0015	0–0.0015	0.002–0.03	0–0.02	0.002–0.03	0–0.002
Ag	0.0001–0.001	0.002–0.2	0.003–0.03	0.01–0.1	0.01–0.2	0–0.02
Sb	0–(0.003)	0–(0.003)	0–(0.003)	0.005–0.1	0.003–0.02	0–0.02
As	0–(0.02)	0–(0.02)	0–(0.02)	0–(0.02)	0.03–0.8	0.1–2.3
Ni	0.001–0.02	0.001–0.02	0–0.02	0–0.02	0–0.01	0.003–0.02
Co	0–(0.003)	0–(0.003)	0–(0.003)	0–(0.003)	0–(0.003)	0–(0.003)
Au	0	0–(0.005)	0–0.01	0–0.005	0–0.01	0–0.01

Table 2:
Chernykh's definition of six chemical groups in the objects from Eneolithic Bulgaria (Chernykh 1978, 79).

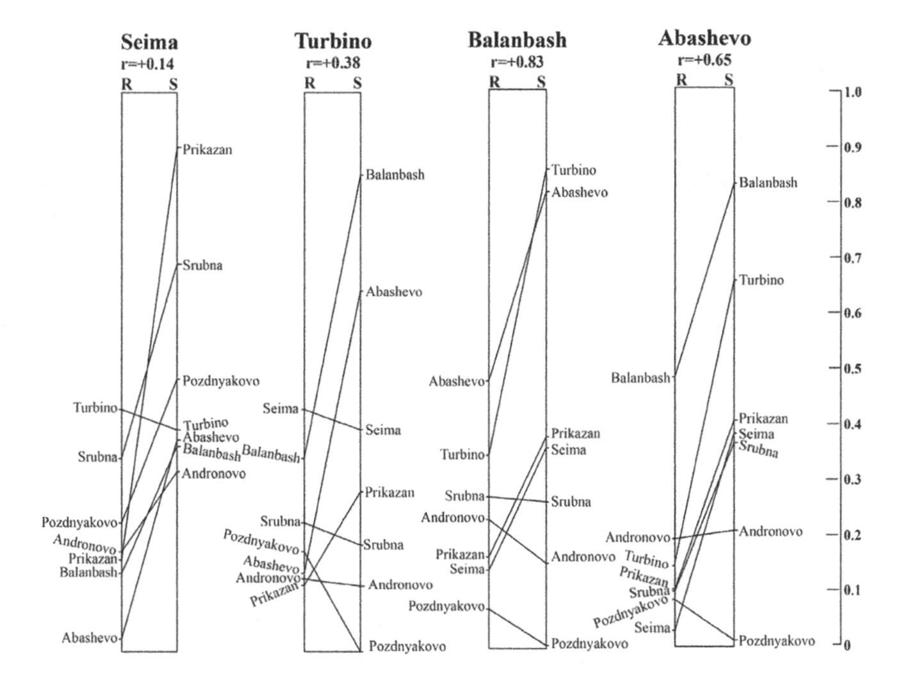

Figure 4:
Part of the comparison of correlations between typological (R) and chemical groups (S) for eight cultural assemblages from the Eurasian steppe and forest zone (Chernykh 1970, 81). Each column represents one cultural assemblage, and the R and S correlations with the other assemblages are marked on the left and right side of each column respectively.

At smaller scales, working within individual cultures or regional traditions of production, he also compared the levels of particular elements in different groups of artefacts to explore functional differentiation in alloy use (e.g., Chernykh 1966, 42). Like Waterbolk and Butler (1962), he focussed heavily on hoards, where these occurred within his study area, and often undertook focussed examinations of their character in order to establish key groupings and better interpret wider regional patterns. In all of his writing, he regularly breaks free from the constraints of his methodology to draw upon his original (and very significant) experience as an archaeologist. Since establishing his laboratory, he has been followed in this tradition by a number of equally dedicated archaeological metallurgists, of whom Sergey Kuzminykh and S.A. Agapov are probably the most widely known outside Russia. Unfortunately, much of their work still remains inaccessible to the majority of European researchers and, in consequence, has rarely been given the credit it deserves. The same is true for the many other Russian and Soviet archaometallurgists who have not been explicitly mentioned here (Bogdanov–Berezovskaya, Degtyareva, Grigioryev, Grishin, Khavrin, Ryndina, Ruzanov, Semilkhanov, Sergeeva and others) and whose work is rarely read outside the CIS or understood in its proper context.

We have chosen to focus at length on the contribution of Evgenij Chernykh, because without his work it would have been impossible to imagine that a complete Eurasian synthesis could be attempted. As we began to understand the details of his approach, it became increasingly clear that, of all of the major analytical programmes described in this chapter, his remains the closest and most comparable, in both methodology and philosophy, to the one we have developed. To quote Kohl (2007, xx):

> "*Although many problems remain unresolved and many paradoxes raised by his work are difficult to ponder, it is impossible to overestimate Evgenij's incredible contribution to our overall understanding of Bronze Age Eurasia. In a sense, we all follow in his footsteps.*"

Barbara Ottaway and the introduction of numerical taxonomy

In the 1970s, Barbara Ottaway was one of the first researchers to work not only with her own analyses, but also to use other published data extensively in order to have a more comprehensive picture of the situation in her study region. Her PhD thesis "*Aspects of the Earliest Copper Metallurgy in the Northern Sub-Alpine Area and its Cultural setting*", published in 1978, was followed by two books, one dedicated to archaeology and society, and the other, "*Earliest Copper Artefacts of the North alpine Region: Their Analysis and Evaluation*", published in 1982, to the analysis of metal artefacts. Her aim was "*to study the earliest metal artefacts in the north alpine region in their cultural context*" (Ottaway 1982, 11). In her own analyses she measured Zn, As, Ag, Sn, Sb and Au with neutron activation analysis (NAA), and Pb, Bi, Ni and Fe with atomic absorption spectroscopy (AAS). Her approach to using the results from different analysts and analytical methodologies, however, raised for the first time the question of data compatibility (see Chapter 3).

Ottaway (1982) was the first to apply cluster analysis and discriminant analysis in a systematic way to the composition of ancient metal artefacts. Cluster analysis is a methodology which uses all the analytical data simultaneously to group artefacts. Typically, each measured element is considered to define a dimension in multivariate space, so that the coordinates of each analysed object in this space are defined by their composition. Thus, if nine elements are measured, an object is defined to be a point in nine dimensional space. Distances between objects in this space can then be calculated using an extension of Euclidean algebra, and thus the proximity of objects to each other in these nine dimensions can be calculated. This is usually followed by a grouping algorithm, which clusters objects together according to the measured proximity in multidimensional space. A feature of this methodology is that it can also be used to combine several different sorts of measurements, such as chemical composition, weight and size dimensions, and even non-numerical data such as shape, providing it is converted into a numerical code. This overall approach to numerical data began in the 1960s under the title of *numerical taxonomy* (e.g., Sneath and Sokal 1973), but the first application in archaeology was by Hodson (1969) when he undertook the analysis of 50 Upper Palaeolithic assemblages of stone tools, and also of a group of 100 chemical analyses of copper and bronze objects, 90 taken from SAM 1 (Junghans *et al.* 1960), and 10 from Schubert and Schubert (1967).

Cluster analysis is a more flexible approach than that used previously by the SAM group, although if new objects are added, then the computer needs to recalculate the similarities between all objects and create new groups. It has also been claimed that the use of a computer to perform the calculations reduces the subjectivity of the analysis, although this is not necessarily the case, since a number of choices have to be made about how the clustering algorithm should work, amongst several other things (see Chapter 2). Since the early work of Ottaway, the methodology of cluster analysis (and related techniques, such as principal components analysis, PCA) has become routine in most archaeometallurgical chemical studies (e.g., Krause 2003; Merkl 2011). Even the critics of such methodologies (e.g., Pollard 1983: see also Chapter 2) do not deny the importance of Barbara Ottaway's work as a milestone in the history of numerical methods applied to archaeometallurgical data.

According to her results on Early Alpine bronzes (Ottaway 1982), the artefacts were initially divided into two groups: bronzes (objects containing more than 2% of tin, an amount that she thought could be related to deliberate alloying) and "copper" artefacts, which have mostly copper with traces of other minor elements (Zn, As, Ag, Sn, Sb, Pb, Bi, Ni, Co, Au, Fe). Cluster analysis identified ten groups (or clusters) of copper, the first of which, representing copper with only small traces of impurities, was further divided into five sub-clusters (see **Figure 5**). The bronzes were divided into six clusters.

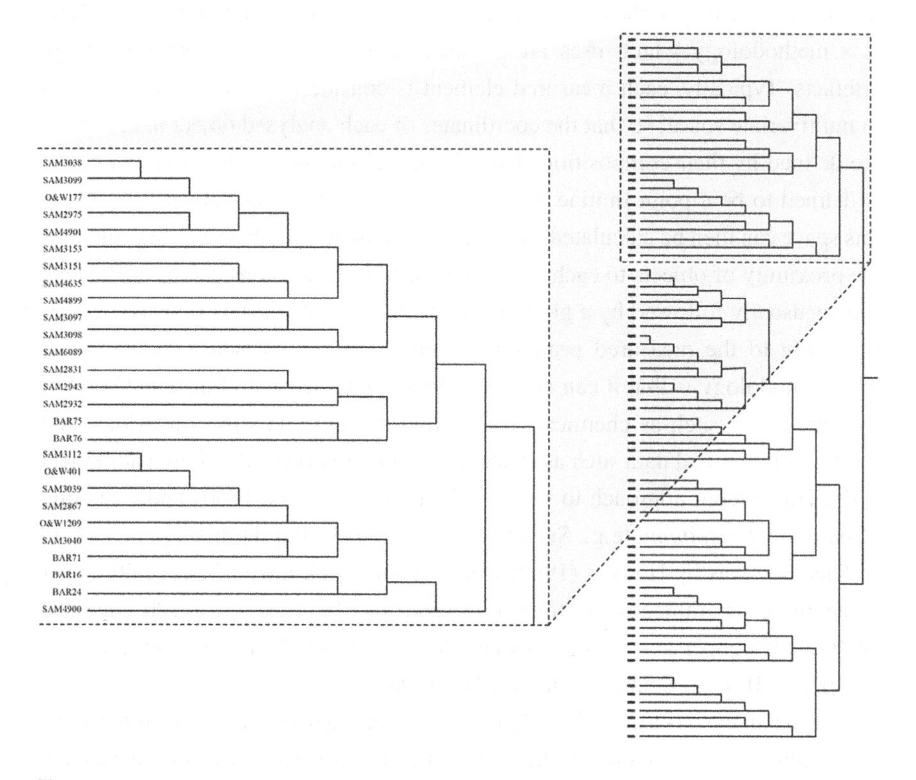

Figure 5:
Section of the output from the cluster analysis of 81 copper objects using 11 elements as input (redrawn and modified after Ottaway 1978).

A further step was to use a numerical approach to verify the correlation between clusters and cultures, clusters and typology, and cultures and typology. She highlighted the relationship between arsenical copper and the earliest cultures (Ottaway 1982, 121–131) and explained this as representing intentional alloying. She also noticed a difference between the compositions of daggers and axes, namely that daggers contained a higher percentage of impurities (Ottaway 1982, 156), opening up a discussion which continues to this day (see Chapter 4). Significantly, she recognized a difference between the copper used for copper objects and that used for bronze production (Ottaway 1982, 156), which the methodology described here allows us to do relatively simply (Chapter 5). She tried to relate different clusters to different copper ores, but in the end she admitted that this was unsuccessful (Ottaway 1982, 171–180). Finally, she posed the question of whether the aim of elemental analysis should be to identify ancient metal ore sources, especially on a large scale, such as the entire European continent. Considering that all the attempts made in the past to do this, including her own, had been largely unsatisfactory, she began a debate about the use of trace elements to determine provenance.

SMAP and Heidelberg

The legacy of the Stuttgart (SAM) group (samples and data) was inherited by Ernst Pernicka's group at the Max Planck Institute in Heidelberg in 1987 (Krause and Pernicka 1996). Attention was focused on Neolithic/Early Bronze Age artefacts, since they were mainly interested in the origin of the Metal Age, particularly on the origin of copper and of tin-bronze production. In particular, Krause dedicated a study to the origin of metallurgy in the zone between the Carpathian basin and Baltic Sea (Krause 2003).

They contributed to expanding the database by adding many new analyses within the *Frühe Metallurgie in Zentralen Mitteleuropa* project (FMZM). They also reconsidered all the work produced by the Stuttgart group within the framework of new ideas and new technologies in the *Stuttgarter Metallanalysen-Project* (SMAP). FMZM focused on eastern Germany, the region formerly known as the German Democratic Republic, which had fallen behind Western Europe in terms of archaeological research. Within this project about 2,400 new analyses were undertaken in Heidelberg using neutron activation analysis (NAA) and X-ray fluorescence (XRF). The analyses undertaken using both methods demonstrated that XRF analyses were not always reliable and, consequently, some of the results obtained with XRF were not included in the SMAP database (Krause 2003, 26). The new analyses obtained within the FMZM project contributed to the creation of the SMAP project, which also included many analyses by other workers. With these new data, the SAM database of 22,000 analyses was expanded to more than 40,000. One aim of this project was the creation of a digital map of all these data. This was very ambitious considering the computers available at the time: neither the computing power nor the GIS packages were then capable of easily managing such a large amount of data.

A fundamental issue raised by SMAP was understanding if and how data obtained from different laboratories using different techniques could be compared, a recurrent topic in the history of archaeometric research since Ottaway's work. As discussed in Chapter 3, the Heidelberg group undertook a study of the reliability of SAM OES data. According to Krause (2003, 18–22), from a comparison of the results obtained by re-analysing the same objects, the analyses undertaken using optical emission spectroscopy (OES) were 'broadly comparable' with modern data in terms of precision and detection limits.

The method of statistical analysis employed in SMAP was cluster analyses, following in the footsteps of Ottaway. The data was based on the elemental concentrations of arsenic (As), antimony (Sb), silver (Ag), nickel (Ni) and bismuth (Bi). These elements were the same as those used by the Stuttgart group, therefore they took the opportunity to re-evaluate the results obtained by the Stuttgart group's statistical analysis, finding that their conclusions were broadly correct (Pernicka 1995, 97). They also considered gold, silver, nickel and—with limitations—cobalt as markers for provenance, and the other elements as characteristic of both provenance and processing. The element pairs silver/nickel, arsenic/antimony and arsenic/tin were chosen to clarify questions of the source of the initial ore and the processing conditions (Krause 2003, 19). Gold was excluded, since in most cases it was below the detection limit with OES (the technique used by SAM). According to the SMAP researchers, recycling of material was probably happening, but this did not change significantly the chemistry of the artefacts and its effect on the cluster analysis was considered as irrelevant (Krause 2003, 145). After statistical analyses, they felt that there were two categories of cluster. One group—"major clusters"—refered to metal that was widespread in Europe, strongly linked to specific cultures (e.g., Corded Ware, Bell Beakers and Únětice) and that signify specific changes at the beginning of the Early Bronze Age. Minor clusters, with a smaller number of artefacts, were seen as expressions of local metallurgical activities.

The work of the Heidelberg group is remarkable for their attempt to reorganize and manage the large amount of data from decades of research on ancient metal compositions. Their idea of integrating data within a GIS package was also valuable and is now achievable with the new technologies available. Their choice of using cluster analysis as the main tool to interpret the data might, however, be suspect in some case where mixing and recycling were prevelant, although the main outcomes outlined above are almost certainly correct (see Chapter 3).

Early Bronze Age	
"A"	Principal impurities: As, Sb, Ag.
	Three Subgroups: "A1" (As>Sb>Ag); "A2" (As<Sb>Ag); " A3" (As~Sb~Ag)
"B"	Principal impurities: As, Ni.
	Three Subgroups:"B1" (As>0.75%, Ni>0.06%); "B3" (As<0.75%, Ni>0.06%); "B4" (As<0.25%, Ni<0.06%)
"C"	No principal impurities. Trace impurities only.
"D"	Principal impurities: As, Sb, Ag, Ni.
	Three Subgroups (based on A1–3): "D1" (As>Sb>Ag); "D2" (As<Sb>Ag); " D3" (As~Sb~Ag)
"E"	Principal impurities: As, Sb, Ni. Subgroups:
	None defined.
"F"	Principal impurities: As, Ag.
	Three Subgroups: "F1" (As>Ag); "F2" (As~ Ag); "F3" (As<Ag)
"G"	Principal impurities: Ni.
	Subgroups: None defined.
Middle Bronze Age	
"M1"	As between 0.65% and 1.05%; Ni between 0.2% and 0.45%; 0.05% Co; trace Sb
"M2"	Similar to "M1" but with As above 1.05%
"N1"	As between 0.35% and 0.70%; Ni between 0.25% and 0.5%; 0.05% Co; 0.05% to 0.15% Sb
"N2"	Similar to "N1" but with Ni above 0.5%
"O"	As between 0.50% and 1.25%; Ni below 0.2%; Sb; Ag around 0.10% and 0.20%
"P"	As between 0.1% and 0.4%; Ni between 0.10% and 0.30%, 0.05% to 0.10% Co; Trace Sb
"R"	Ni only principal impurity
Late Bronze Age	
"S"	Principal impurities are: As, Sb, Ni, Ag. Both As and Sb are generally over 0.40 to 0.50% with 0.25% Ni and 0.25% Ag. Sub-groups are defined by the ratios of Sb to As. "S1" with Sb:As = 1:1; "S2" with Sb:As = 2:1
"T"	Similar in character to "S" but with lower levels of impurities, all below 0.40%

Table 3:
Northover's chemical classifications of Early, Middle and Late Bronze Age European metals (Northover 1980, 230–231).

One factor contributing to the lack of consistency between the results of all of these projects is that each group of researchers derived their own methodology for separating out 'chemical groups'—initially variations on the themes of 'decision trees', with break-points for particular elements calculated from the distributions of each element, and latterly based on cluster analysis. A system that combines the most consistent features of all of these previous 'decision tree' type studies is that developed by Northover (1980: see **Table 3**), which has become the most widely used classificatory system in recent western European metallurgy. Noting the issues over interpretational differences between major analytical programmes, Ernst Pernicka, one of the leading archaeometallurgists of the last few decades, has concluded that this lack of consistency is due to insufficient attention being given to the geochemistry and metallurgical behaviour of the trace elements. He asserted that by applying careful consideration to these factors, reliable provenance determinations can be made (Pernicka 1999). He noted the lack of consensus about what conclusions archaeologists can draw from these major European programs of chemical analysis of Bronze Age metalwork, and stated that "*[T]race element analysis of ancient metal objects was so discredited that for some time hardly anybody looked at the data available*" (Pernicka 1990, 169). He emphasised that, for sound geochemical reasons, not all trace elements can be taken as reliable indicators of provenance, and concluded (as Chernykh had done in the 1960s) that only those which follow copper in the smelting process (as opposed to concentrating in the slag phase) are likely to be reliable in provenance studies. The most widely reported of these elements are As, Sb, Ni, Ag and Bi, although even some of these can also be partially affected by subsequent technological processing. He also considered the volatility of some of these elements, and noted that when smelting sulfosalts from Cabrières in southern France, primarily tetrahedrite ($(Cu,Fe)_{12}Sb_4S_{13}$) but with some tennantite ($Cu_{12}As_4S_{13}$), the Sb was significantly lost during smelting, whereas As, which is more volatile under oxidation than Sb, was lost proportionately less because it is present at lower levels.

Lead isotopes

Largely because of the growing scepticism concerning the use of trace elements for determining the provenance of copper alloy objects, during the 1980s archaeologists enthusiastically adopted the newly developing technique of lead isotope analysis. Brill and Wampler (1967) had shown that it was possible to differentiate the lead from Laurion in Greece from that obtained in England and Spain by using measurements of the lead isotope ratios. They did note, however, that an ore sample from north-eastern Turkey fell into the same 'isotope space' as that occupied by three ores from England, thus suggesting that not all ore deposits had unique isotopic signatures. The scope for using lead isotopes was widened once it was realised that not only could it be applied to metallic lead artefacts (which are, archaeologically speaking, rare), but also to the traces of lead left in silver objects extracted from argentiferous lead ores by cupellation (Barnes *et al.* 1974). However, its full potential was subsequently developed following the discovery that it could also be applied to the traces of lead remaining in copper objects smelted from impure copper ores (Gale and Stos-Gale 1982). This latter discovery potentially provided archaeometallurgists with a powerful new tool to provenance copper alloys, but unfortunately it became somewhat bogged down in controversies over the interpretation of the data during the 1990s (see, for example, Pollard 2009). A re-thinking of the use of lead isotopes in archaeological copper alloys is presented in Chapter 6.

Beyond Provenance?

The title of this volume is intended to be provocative, and is aimed at encouraging archaeometallurgists to think beyond the simple question of 'provenance'—taken here to mean 'where does this metal originally come from?'—and to focus more on the causes of change within the archaeometallurgical record, which includes provenance but goes much wider. As shown above, the chemical analysis of archaeological copper alloy objects began in Europe more than two hundred years ago, and there have been several major analytical programmes of European Bronze Age metalwork during the 20th century. There can be no doubt about the immense contribution made by all of these projects and others, not least of which is the publication of the chemical analyses of at least 50,000 Bronze Age metal objects, mostly from Europe, but with a significant number from the Caucasus, the Urals and Central Asia. Nevertheless, this rich data legacy also presents us with several challenges. Can we afford to ignore such a large corpus of data, simply because it does not conform to modern standards of analytical quality? We could choose to restrict ourselves only to data produced after, let us say, the year 2000, using methods such as the electron microprobe or inductively coupled plasma–mass spectrometry (ICP–MS), but to do so would be to severely reduce the volume and geographical spread of the available data. An alternative, and the approach taken here, would be to develop an interpretative protocol which can accommodate the obvious shortcomings in some of the data. This vast legacy of data, without which we could not have begun to undertake the project which underpins this volume, provides the foundation for the methodological developments presented here. The original interpretations of these data, although perhaps now in need of review, have set the benchmark for the development of modern archaeometallurgy, and have provided insights about the exploitation patterns of copper in the Bronze Age which continue to be used today (e.g., Thornton and Roberts 2014).

In a recent valuable summary of the current state of affairs in European archaeometallurgy, Radivojević *et al.* (2018) demonstrate the deep hold that the concept of provenance still has on the minds of many European archaeometallurgists. This is most clearly seen in the conclusion, which states that the purpose of archaeometallurgical research has been to "address archaeological questions of alloy selection, development, distribution, and provenance, the latter long considered the 'Holy Grail' of the discipline." Despite the generally negative view, as expressed by Ottaway and others, of the usefulness of chemically provenancing metals to specific ore sources when it is applied as a "black box" technique, there can be no doubt that when focussed on a specific region and carried out in combination with a detailed holistic view of the ore mineralogy and metal smelting debris, provenance studies can produce significant results (e.g., Stöllner and Samašev 2013). To be done well, however, such work is extremely time-consuming, since it requires considerable archaeological and geological fieldwork, combined with a very large number of chemical, metallographic and petrological analyses. We outline a different option, as explained in Chapter 2, which is to suggest that in some circumstances provenance may not be the only, nor perhaps the most meaningful, archaeological question to be asking.

Radivojević *et al.* (2018) also point to the increasing practice of hoarding metal in Europe from the mid-second millennium BCE onwards, and associate this with increased metal recycling. They note that: "*how to recognize recycling and determine when and the degree to which it occurred is a promising research area*". Many of the previous studies (with the notable exception of Chernykh) have either ignored completely the potential complexities arising from the mixing or recycling of metal from different sources, or have acknowledged that such practices would invalidate the simple hypothesis of provenance, but have then largely dismissed the problem. Crucially, what is not yet clear is the balance between situations where traditional provenance is meaningful and achievable, and situations where, because of intense recycling, it is not. We argue in Chapter 2 that this balance is likely to be different over time and place, and even between classes of object. In other words, there will be circumstances in which a 'traditional' approach to provenance is worthwhile, and those where it is not. We have focussed our attention on developing methodologies to address this latter situation, but we should note that the methods developed here to consider such complex circumstances work equally well in situations where provenance is meaningful, and therefore nothing is lost by taking such an approach. Furthermore, given the plethora of mutually exclusive interpretational methods applied to chemical data in the past, another of our objectives has been to provide a 'universal' system for approaching the initial classification of chemical data. This is not to imply that the *archaeological* contexts and interpretations are universally similar, but merely to say that the initial methodological approach to the data is the same wherever the data originate, and that new data can be added without requiring complete recalculation of the results.

Archaeology has progressed much since the launch of the major analytical programmes on copper alloy artefacts in the 1930s and 1940s, and certainly since the idea of determining the provenance of archaeological artefacts by chemical means was first articulated in the 1860s. Our questions have evolved from the apparently simple, but in fact quite complex, one of "*whether there are any means of recognizing the locality from which the metal in a given copper object was obtained*" (Coghlan *et al.* 1949), to a set of more complicated and socially-embedded questions about how humans actually *used* and *circulated* metal objects (see, for example, Kienlen 2013). As summarized by Radivojević *et al.* (2018), modern views on the history of metal use have evolved considerably from the early assumptions driven by either technological or geological determinism. The older vision of bronze reflecting a static sequence of 'metal industries', as articulated by Childe and others, has been replaced by ideas of a "*dynamic 'metallurgical landscape' during the European Bronze Age, with numerous local and regional metal producers feeding the demand for metal*" (Radivojević *et al.* 2018).

Our approach, as explained in Chapter 2, is an attempt to visualize this much more fluid world. It entails understanding not only the life history (*object biography*) of individual objects, but also emphasizing the characteristics of *assemblages* of objects, and consequently the life history of the flows of metal from which these objects were made. To use a simple analogy, if we think of the extraction of metal and the manufacture of an artefact, then this is merely the 'birth' of the artefact, which might subsequently go on to have a long and varied life history. A focus on provenance then becomes analogous to thinking that the only factor of importance in a person's life is where he or she was born. We aim to show that the chemical and isotopic composition of assemblages of objects contain patterning which can potentially reveal not only the 'birthplace' of such objects (i.e., the provenance), but also other significant events in their life history. Sometimes these events might render the determination of provenance difficult if not impossible, if they involve the mixing of metal from more than one source, or the recycling of objects. Nevertheless, we would argue that valuable information (and indeed, archaeologically speaking, information potentially even more valuable than provenance) can be derived from a chemical study of the life history of these objects, which is, of course, a history of the interaction between objects and humans. Put simply, we believe that the focus of the interpretation of chemical and isotopic data from archaeological copper alloys should start with detecting *change* in the material record, rather than concentrating solely on determining the *source* of the metal. After all, the basic methodology of archaeology is to detect and interpret change in the archaeological record. Our aim here is to show how we might build upon and add to the achievements made during the earlier phases of archaeometallurgical study, by developing a new set of methodologies based on these ideas, which ultimately aim to put the use of metal more firmly into its social context.

Chapter 2

Developing a New Interpretative Framework

Perhaps the most important characteristic of the framework and the consequent toolkit we have created for FLAME is that it focusses on detecting and quantifying *change over time* and *differences over space* in the *archaeological* record. In our case, the key elements in the archaeological record are the metal objects themselves, and our aim is therefore to produce a toolkit which reveals observable changes in these objects over time and space. Our entry point to identifying these changes is via the chemical and isotopic composition of the objects themselves, but such changes can only be meaningfully interpreted when contextualized by their typology, decoration, archaeological context, and manufacturing technology. The ultimate aim of FLAME is to use the changes revealed by a comparison of the chemical and isotopic data to infer human action and intention. It is this focus on 'change', and the explicit intention to interpret such change in terms of human action, that distinguishes our philosophy from the more limited ambitions of some of the previous analytical projects reviewed in Chapter 1, focussing primarily on 'provenance'.

Such inferences, however, cannot meaningfully be made on the basis of individual analyses on isolated objects. These analyses give no sense of what is 'normal' or 'expected' for a particular object type in a specific time and place, and therefore lack any comparative context—this is effectively the same argument as that made by Pittioni (1957) when challenging the work of Otto and Witter (1952), as discussed in the previous chapter. Observations of change and difference have to be made on the basis of *group properties*, which should be determined on as large a number of objects as possible. We term such a set of objects an *assemblage*. A variety of views on the meaning of the term 'assemblage' have recently been published in a special issue of the *Cambridge Archaeological Journal* (2017(1)), where, as with many commonly-used archaeological terms, one finds that the concept itself is fluid. As discussed further below, we see an assemblage essentially as a thematically-defined group, the nature of which is not fixed but depends on the specific question being asked. In our case, an assemblage could be all the metal artefacts from a particular tomb, or all the metal objects belonging to a particular archaeological culture, or all the bronze daggers of a particular shape from across Eurasia. The toolbox described in the next chapters is designed to reveal changes in chemical and isotopic composition through a comparison of such assemblages.

One other key point to note is that our attention is intentionally focussed on *objects*, or more strictly on *assemblages of objects*, as opposed to other sources of metallurgical data, and our methodologies have been developed accordingly. That is not to say that we are not interested in these other sources of evidence relevant to metal artefacts, such as the mineralogy and chemistry of known ancient mining sites (as provided by, for example, O'Brien 2015), or the chemical metallurgy of the smelting process, or the metallographic structure of the objects themselves. However, we regard these as *independent* sources of information, to be subsequently compared with the results of the analysis of the objects themselves. Thus, if we identify a particular region as being dominated by objects made from a particular type of copper (see Chapter 4) in a specific time period, then this offers us a starting point to postulate that a mine or mines operating at that time and producing such copper should exist within that particular region. The identification of a mine capable of doing so then provides *independent* evidence that such copper was being locally produced. It is however worth noting that our mapping approach for different types of copper can also suggest the geographical extent of the use of such copper, which is information not immediately obtainable from the examination of the mining area itself. Conversely, the absence of such a suitable production area within a particular region would strongly suggest the importation of metal from another area. The use of a GIS database allows for the overlaying of several such independent sources of information, thereby facilitating the combination of these multiple strands of evidence.

Building a new conceptual framework: 'Form and Flow'

Our conceptual framework is the major feature that distinguishes our approach from those of many previous workers, but to some it may at first glance appear unnecessarily abstract. In our view, however, it provides a powerful new framework which allows us to combine data from many different sources, and in particular to link chemical and isotopic data to human behaviour (Bray *et al.* 2015). As such, it has been very helpful in guiding the development of the tools and ideas discussed below. Rather than focussing on specific metal objects, the central concept is one of metal 'flowing' over time through society, and being chemically and isotopically modified by a series of human interventions, which

will in turn influence the composition of new objects produced from that flow. Such interventions might include:
- the mixing of ore or smelted copper from more than one mining source;
- deliberately alloying copper with significant quantities of another metal, such as tin or lead, to create a new material;
- re-working an object into a new shape, possibly with re-alloying and/or the addition of other metal, or recycling objects to create new objects.

We can see that many such factors can combine in a wide variety of ways to produce a complex dynamic system which affects the composition of the flow of metal and the metal artefacts made from it. Our aim has been to develop a quantitative methodology to follow and disentangle this system. It should be immediately obvious that it automatically includes the traditional concept of provenance, if the situation is simple enough to do so, but is also capable of dealing with more complex scenarios.

Metal flow

The concept of metal flow in archaeology has been emphasized by several scholars (e.g., Bradley 1988; Needham 1998; Jin 2008; Pollard 2009), often in the context of attempting to model trading networks and technological pathways, or to express the life cycle of an object as it is made, used, and deposited. An example of the latter is illustrated in **Figure 1**, from Ottaway (2001), in which the metal is seen to go around a cycle starting with mining and smelting, through the manufacture and use of the object, and ultimately back to an oxidised form through corrosion. It is effectively a thermodynamic cycle in which energy is consumed in converting the ore to the metal, and entropy eventually causes it to revert to its mineral state. Such a cycle is a useful representation of the practical cycling of metal, but our 'flow', dealing with the underlying metal, is more abstract than this (Bray and Pollard 2012). Essentially it is a theoretical construct which enables us to separate the lives of individual objects from that of the metal from which they are made, and ultimately to link together the data from mines and smelted metals to objects.

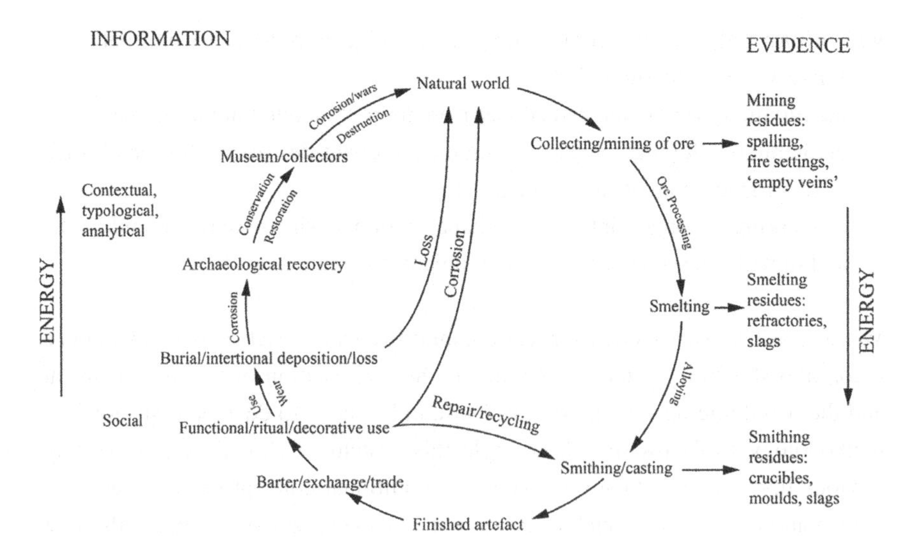

Figure 1:
Cycle of copper production and working (redrawn after Ottaway 2001).

In creating this model, we have relied heavily on the theoretical construct of the 'biography' of an object (Gosden and Marshall 1999). The concept of a person's biography is well understood and well established. It is a record of a person's birth, life, deeds and death. It examines the way in which a person interacted with the world, and the way that the world interacted with them. Just as every person has their own unique biography, so too does every object. At the heart of this notion of object biography are questions about the links between people and things: about the ways that meanings and values are accumulated and transformed. Biography is relational, and an object biography consists of the sum of the relationships from which it is created. As an object goes through the course of its functional life it interacts with people. Such objects do not simply set the stage for human action, they are integral to it. In our case, however, we need to think about the distinction between the biography of a single object and that of the underlying metal flow. A specific object may have only a relatively brief existence, but the metal flow from which it is made, and to which it might be returned if it is recycled, may have a much longer existence. In many ways this concept is similar to that of *prosopography*, familiar to historians. This describes the situation where each individual within a defined group, such as all the bakers in Medieval Nottingham, might have left a relatively sparse biography (e.g., birth date, marriage date, location of bakery, etc.). However, when all these sparse records are assembled and integrated, we may get a good picture of the life of bakers in Medieval Nottingham—not of any particular individual, but an average of the lives of this *assemblage* of bakers. Our metal flow is in some ways equivalent to the assemblage of bakers' lives—an *object prosopography*.

Simple linear biographies

The biography of a single object may be conceptually very simple—it could be made, used for a short period, and deposited, to be later found by archaeologists. We term such an object biography a *simple* or *linear trajectory*. As a trivial example, we may imagine that a member of the elite demands a particular object be made using primary copper from a specific source. We use the term 'primary' here to denote copper fresh from the smelter or refinery, and not mixed with copper from any other source. This primary unit of copper is probably then alloyed with tin and perhaps lead to give certain desired physical or visual properties. The object is made from this metal stock and performs a specific function for a period of time in the elite household, and is then buried within the tomb of the person, probably only a few years or decades after manufacture. In this simple scenario, the object then sits in the tomb until it is excavated, conserved, chemically analysed, and perhaps put on display in a museum. Although from a theoretical perspective we must also consider the period in the tomb and on museum display to be part of the object's biography (van der Stok-Nienhuis 2017), in terms of our flow model the object is only part of the flow of metal for the period when it is in active use, from manufacture to deposition. After that, it is effectively 'out of circulation' and not able to contribute to, or be affected by, the flow. Moreover, even when it is 'in circulation', in this particular case it undergoes no significant changes—after it has been smelted, alloyed, and manufactured, neither its composition, form nor decorative features are altered until after it is removed from circulation. Essentially it carries the same information into the grave as it had when it was first made. In this case, it is an instantiation of the composition of the metal flow available to the metalworkers at the time it was made. For objects which follow such a pathway, the traditional chemical and isotopic approaches to provenance are likely to be feasible, and possibly successful (**Figure 2**). Any 'fingerprint' inherited from the ore source is highly likely to be preserved within the object. The traditional 'provenance' models using chemical and isotopic data appear to have generally assumed (often only implicitly) that *all* archaeological copper alloy objects more or less follow such simple (linear) paths. Our view is that this may be the case, but the onus is on the archaeologist and analyst to explicitly demonstrate that this is true *before* moving to undertake provenance studies.

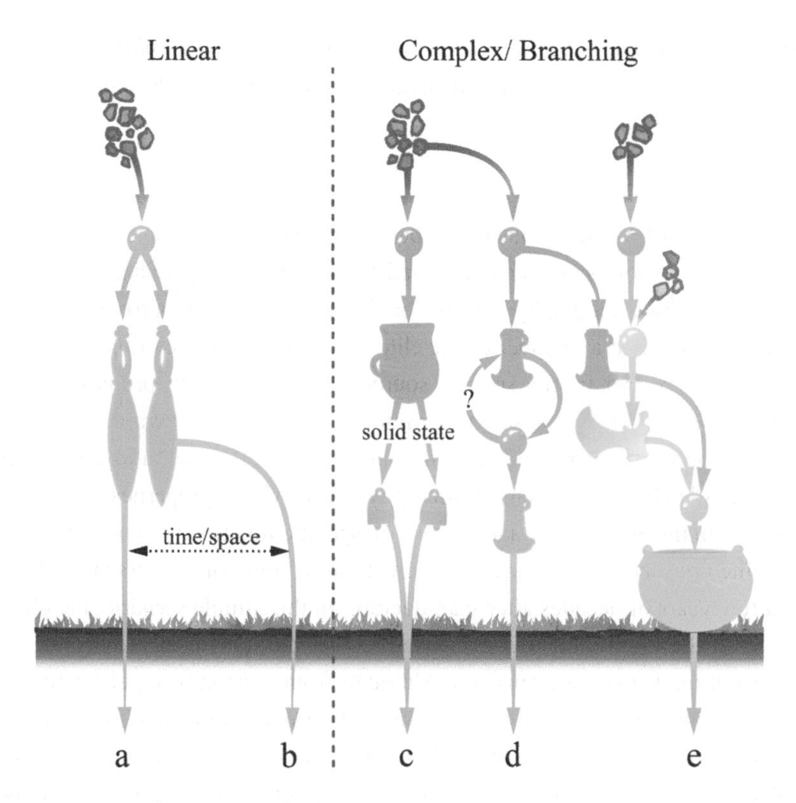

Figure 2:
Linear and complex object biographies.

Complex biographies

Although such a simple short biographical pathway between origin and deposition is of course possible, it is obviously not the only trajectory that we might imagine. For example, instead of being buried in the elite tomb, the object could have been passed on to succeeding generations, either as a practical object for further use, or as a memento, or an heirloom, to be buried some time later with a descendant of the original elite person. It might thus remain in use for several generations, being inherited, curated, and passed along repeatedly. This scenario simply extends the active life of the object. Following burial, however, it might have been looted and re-used in its original form, but in another time and, possibly, place. This too would lengthen the 'active' life of the object, which would then have had an 'interrupted' life history, both in time and perhaps also in space, if it is transported some distance after looting. All of these possible combinations of scenarios simply extend the 'active' life of the object, but do not of themselves physically change the object, nor the chemical information contained within it. As in the previous example, it still carries to the tomb the same information that it had on creation, and is still an instantiation of the flow of copper at the time it was made. If, however, it was looted from a tomb, or passed on into a new social context by some other mechanism, such as trade, exchange or gift-giving, it may no longer have had the meaning that it did in its original context, and might have been melted down to create one or more new objects. The original object, in this new social context, may have been seen more as an *ingot* containing a convenient supply of raw material rather than as an *object* containing significant symbolic capital. At this point it may simply have become 'scrap metal', to be mixed with other unvalued object forms, and potentially reworked into other completely unrelated forms. We may imagine that such processes could continue in this way for some time, until the object is finally lost or deposited. We would term such a life history a *branched* or *complex biography*. The mutability of copper and its relative resistance to corrosion lends itself to such long and complex lifetimes, although perhaps not to the same extent as gold and silver, or possibly even glass.

But, in this last scenario, to which object or objects are we actually attaching this biography? It cannot be to the original object, unless we choose to see such a chain of events as being composed of a sequence of related objects, each with their own biographies. This might be appropriate for a particular set of events, where a single object is re-made at intervals, but with no addition of new metal. However, at each re-melting event in such a chain, there is the potential for the metal from one object to be divided between many objects, or for many objects to be amalgamated into one object, or new metal from a different source to be added. Under such circumstances, we suggest that it is better to switch our focus from *the biography of individual objects* or a sequence of objects to the *biography of the metal* contained within these objects, since it is this metal which is actually being manipulated by human agency. This is our conceptual flow of metal. It 'flows' through the objects, but its composition can change over time, as new sources of metal are added to the flow, even though the composition of each individual object within the flow may not change over its own lifetime. Moreover, we can conceive of a metal flow which can change composition over time without objects being recycled, simply by a new stock of metal being injected into the flow. This new stock may come from a new mine source being added to an existing flow of metal from the original mine, causing a significant change in the composition of the flow. It could also be that the flow is interrupted—metal from one mining region becomes no longer available, and is simply replaced with material from another source, with no continuity between the two. Hence the flow model is not predicated solely on the recycling of objects, but takes into account the multiplicity of events which might befall the metal flow.

To aid the explanation of this model, in a previous paper (Bray *et al.* 2015) we likened the flow of copper to that of a river, but this analogy is perhaps unhelpful if taken too literally. There are certain features which work, such as the idea that objects can be 'scooped out' of the metal flow like water in a bucket, and returned to the river if recycled. The concept of tributaries joining together is also a useful analogy for multiple mines providing metal to the flow, but the fact that rivers tend to grow in volume from source to sea does not necessarily apply to metal flow. It is more likely that the flow is greatest nearer to the source(s). In that paper, we also placed great emphasis on the role of recycling of individual objects or groups of objects in changing the composition of the flow. This was perhaps useful in the context of the Early Bronze Age in Western Europe, where we think that down-the-line trade and recycling may have been part of the dominant mode of metal transport, as a consequence of objects being passed between groups of people over relatively short distances, and re-modelled to fit local expectations of shape. More generally, however, we feel that in many cases this process is likely to have been a minor contributory factor compared to the greater volumes of raw metal being injected into the flow from new mining and smelting sites. We therefore see this hypothetical 'flow' of metal as being a useful tool for linking the composition of metal flowing from many mines, as well as being a mechanism for handling the possibility of the large-scale recycling of objects. It is almost certain that the balance between the influences of these two mechanisms will vary over space and time, as well as by the form of the object (perhaps weapons being treated differently to more mundane objects), and also the social status of the potential users of the objects. Nevertheless, by developing a series of tools which allows us to detect change in the flow of this metal, we believe we provided a practical framework for identifying and untangling the complex nature of the interaction between humans and metal.

Flow and provenance

It is worth reflecting at this point what the differences are between a dynamic 'flow' model as described here, and other more traditional scenarios, since this is a key distinguishing feature of the FLAME project. In the case of an object with a simple linear biography, or any linear biography in which the original object remains intact, then the conventional 'provenance hypothesis' as first enunciated by Damour (1865, 6) and summarised by Wilson and Pollard (2001) clearly applies. There is likely to be some characteristic of the ore that, after allowing for the changes which can occur in smelting and manufacture, is carried through into the object. This can be measured and, after comparison with appropriate ore data, or some other material of known origin, can be used to assign an object to an ore source. Strictly speaking, of course, such a procedure can only *eliminate* sources from which the object could not have come, rather than *prove* the metal to have come from a particular site. Nevertheless, the ambition of provenance can in principle be achieved. But what of the second scenario? If metal from many ore sources is mixed, as might be necessary in large-scale bronze production, or if several objects are recycled to make something new, then the simple link between a single ore source and the object is gradually destroyed. Ultimately, an object ceases to have a single source, and the simple 'provenance hypothesis' becomes meaningless.

Figure 3 shows a series of schematic interpretations of some of these scenarios. The first sketch shows a set of objects recovered at different times, and a series of inputs, which probably represent different mining sites. Between them is the hypothetical metal flow, which consists of all the objects (plus scrap metal, ingots, etc.) that exist at a particular time and place. The composition of this flow is of course unknown, both in terms of the total population and typology of the objects, and also its range of chemistries. The second sketch illustrates how we can use the chemistry of an assemblage (the recovered and analysed sample) to approximate the chemistry of the metal flow at that time, providing we can be satisfied that the assemblage is sufficiently representative of the parent population—the flow. The third suggests how separate metal flows might co-exist—in this example, an elite flow and a common flow, each drawing on different sources of metal, and giving rise to different assemblages. This sketch also indicates a fracture in the metal flow caused by a culture change, which we hypothesise, changes the metal supply systems, such that the composition of the flow changes. There is, however, the possibility of some continuity across the transition, if the later culture robbed or reused metal that was in circulation before the change.

A key question, and one which we may never be able to answer satisfactorily, is '*how many archaeological copper objects conform to the single source hypothesis, and how many are too complex for this to be a meaningful question?*' We can postulate some general answers to such a question, which may or may not be helpful. In regions close to a single large mining source, where fresh metal supply is plentiful, we might expect there to be little mixing or recycling, and therefore the provenance hypothesis is likely to be valid. Further, we may expect the degree of mixing/recycling to increase with distance from such a source (although distance need not be a linear measure, but may be directional, and depend on factors such as ease of river transport, etc.). Such a simple linear relationship might also apply but in a different way to a highly organised complex society, where the means and resources exist to transport large quantities of metal from very specific sources

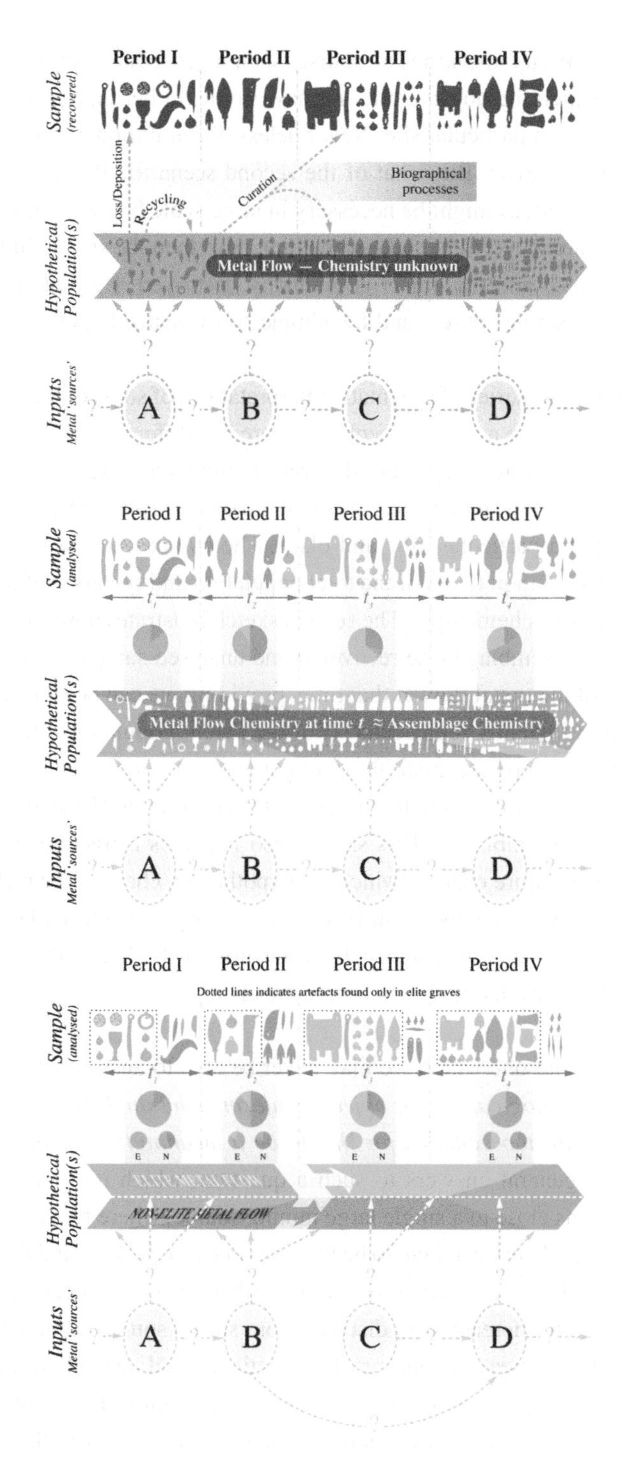

Figure 3:
Hypothetical illustrations of metal flow.

over considerable distances. The reverse could, however, also be true—highly organised complex societies might have the resources to draw in metal from many sources, and mix them at the foundry sites, thus negating the provenance hypothesis. In metal-consuming regions which are remote from metal sources, where the supply may be difficult or discontinuous, we may expect a far higher degree of mixing and recycling. Geography may not, of course, be the only factor. Political interventions (or, indeed, natural disasters) might disrupt the supply from a particular region, causing a switch of the inputs into the flow. Societies in terminal collapse, where central control has diminished and they no longer have the capacity to control the metal circulation, may also see a rapid change in the balance between the use of fresh metal and recycled metal. We see just such a situation at the end of the Roman occupation of Britain, where we have postulated that by the end of the Early Saxon period, some 250 years after the collapse of Roman Britain, at least 75% of the copper alloy in circulation was still recycled Roman (Pollard *et al.* 2015).

However, under the flow model, all is not lost in these more complex scenarios, even when mixing and/or recycling may become significant. In such situations, we must switch to a stronger focus on thinking about *detecting change* in the flow of archaeological metal using the appropriate tools, rather than simple provenance. This means that the traditional question of provenance has to take a back seat, although it is not forgotten altogether. It merely becomes one of a number of possible explanations for the observed chemical changes in the metal flow. In other words, the cause of these changes may not be a simple switch in ore source, but might reflect many other factors, such as a massive input of looted metal, or a change in smelting practice, or a general increase in recycling. Nevertheless, we suggest that the observation of such changes within the metal supply is still archaeologically meaningful. In fact, we might argue that the ability to detect such changes is more archaeologically meaningful, and possibly more interesting, than simply thinking about changes in ore source. Put more philosophically, we might argue that our model attempts to deal with the general case of human interaction with metal, whereas provenance is an example of a specific case, sitting within the more general and potentially more complex situation.

Operationalizing the framework

Several groups of researchers have articulated frameworks for understanding metal circulation which are similar in many respects to that described above (in particular, Chernykh (1992) and Ottaway (1982)). Our model differs primarily by explicitly separating the biography of the individual objects from the conceptualised biography of the underlying flow of metal. But what in reality is this 'flow' of copper? It consists of a series of snapshots of the stock of copper (or copper alloy) available at any particular time and place, and the 'flow' reflects how this stock changes over time and space. The stock is made up of all the available copper alloy resources at that time and place, including all objects, plus fresh metal 'ingots' and other 'scrap', and the chemical and isotopic composition of the stock is therefore the average of that in all of its components. However, its precise composition is, of course, generally unknown, and largely unknowable, to us. We can, however, tentatively make the assumption that the composition of those objects that are available to us (i.e., all the chemical and isotopic analyses of objects from that particular time and place, which is one definition of an *assemblage*) is a *representative* sample of the composition of the stock. This allows us to reconstruct the chemical and isotopic composition of the metal in circulation at that time and place, and, hence, to compare the metal in circulation at different times or places. This assumption is of course only true providing the data available to us are an unbiased sample. As is inherent in all aspects of archaeology, this is unlikely to be completely true, perhaps through the vagaries of sample preservation, or excavation strategy, or through biases in the selection of samples for analysis (arising from museum sampling constraints, interests of the analyst, etc.). The issue of sample bias is addressed in more detail below.

The toolkit: the "Oxford system"

Based on the model developed above, we have devised a system (sometimes referred to as 'the Oxford system'), which is based on a set of three separate but interlinked groups of tools:

- **trace element composition or 'Copper Groups'**, which focusses on information derived primarily from the copper ore source(s), but which may potentially be altered by subsequent human manipulation of the metal,
- **alloy composition ('Alloy type')**, which is defined to be the result of intentional action, as craftspeople choose to add metals to modify the characteristics of the material (fluidity in casting, colour, hardness, etc., or perhaps to give additional symbolic significance), but subsequent mixing and recycling might move the assemblage away from the originally-designed alloy compositions,
- **lead isotope composition,** which can give information about the source of copper, or the added lead, but is also susceptible to alteration due to anthropogenic mixing.

These tools are based on the intrinsic properties of the objects (i.e., on their chemical and isotopic compositions), to which we add at least one more intrinsic property, that of **form** (as described by typology), which is imposed by humans and reflects the socio-technological context of production, amongst other things. We can of course elaborate on this, by considering form to be a complex variable incorporating decoration, manufacturing technology, the 'technological style' of Lechtman (1977), etc. There is a fifth (extrinsic) property, namely '**context**', which frames the life history of each object, and allows us to situate its intrinsic attributes within the wider physical and social world.

It is important to note that, in general, and if the data support it, we would normally attempt to use all three of these toolkits together to address a particular archaeological question. Frequently this is not possible because we cannot get a full set of major and trace element data, plus isotopic measurements, on the same objects. If it is possible, however, then it is generally worth doing because each toolkit provides a specific perspective on different aspects of human behaviour, and it is only by piecing them all together that we can hope to obtain a realistic answer. In general, the trace elements contain information about the source (or sources) of the copper, and might provide some evidence for the manipulation and recycling of the metal. The alloying data gives us a picture of the desired alloy composition, if it is a deliberate alloy, or allows us to demonstrate that the assemblage represents a set of objects which have no common target composition, which might be indicative of extensive recycling. The toolkit is therefore deliberately designed to consider the identity of the copper in circulation separately from any alloying processes. This separation is a unique and powerful feature of the Flow model. Alloying can be a deliberate choice, aimed at producing a metal with specific physical or aesthetic properties, but equally it may be a series of deliberate choices, where a particular alloy may be *re-alloyed* sometime during its lifetime, perhaps by adding more tin to change the colour, or by adding lead to increase the fluidity of the melt, or to dilute and extend the stock of metal. We do not therefore see alloying as necessarily being a single unique event in the life of the flow of metal. Thus, neither the percentage of the alloying elements present in an object, nor the consequent 'alloy type', is a fixed property within the metal flow.

In modern foundry practice, and presumably also in ancient practice, it is not unusual to sort alloys before recycling, so that the composition of the final product can still be controlled to some extent. However, in a situation where recycling is extensive and/or not very selective, we may also envisage that the make-up of the alloy becomes increasingly less controlled, and ultimately results in metal where the alloying elements are present at more-or-less random levels, which may be below the levels at which the alloying elements exert much influence on the physical and aesthetic properties. In contrast to the modern (engineering) definitions of alloys, our approach is specifically designed to highlight such situations, and distinguish between intentional 'primary' alloying and less intentional 'secondary' alloys.

In particular, we can consider circumstances where some copper from a specific source (or mixture of sources) is circulated in an unalloyed form, whereas some of the same copper is alloyed and then circulated in the alloyed form. By conceptually separating the processes of producing the copper and producing the alloy, we can therefore begin to ask questions about the form in which the metal was circulated—perhaps as separate ingots of copper, tin and lead, to be subsequently mixed to order at the foundry, or as preformed ingots of copper with tin (or lead) already added. The practice of recycling would suggest that the raw material at the foundry might also include objects which have no specific significance or value. This in turn prompts a discussion of what the commonly-used term *ingot* actually means. Normally it is defined as a block of raw single metals (i.e., ingots of copper, tin and lead) transported specifically to be used at a foundry to create objects. We believe that we must allow for the possibility of ingots consisting of pre-alloyed metals, but also that, in some circumstances, an unwanted object may simply be regarded as an ingot of raw material. In other words, one person's axe might become another person's 'ingot', and this might be an important mode of trade and exchange in some contexts. It does not require much imagination to see that in some circumstances recycling other people's metal objects is likely to be far easier than mining, smelting and transporting fresh metal. This leads to the introduction in Chapter 5 of the idea of *regional alloying practice*, where we might look for regional patterns in how alloys were designed and produced.

In Chapter 6, we present new ways of presenting lead isotope data, which differ from the conventional approach of plotting a scattergram of two sets of isotope ratios. The purpose of this is, however, more than simply to explore new presentational techniques: it represents a fundamental re-think of the use of lead isotopes in archaeology (Pollard and Bray 2015). The conventional approach simply represents the adoption of the interpretational techniques developed in lead isotope geochemistry, the original purpose of which was to provide a graphical means for calculating the geological age of particular lead deposits. Although useful in some circumstances, we suggest that the interpretation of lead isotopes in *archaeological objects* is different from that in *geological ores*, primarily because of human action—the possibility of mixing lead from different sources, the addition of lead to copper objects, or the recycling of objects containing lead from one source into the flow of metal which might contain lead from other sources. In other words, there is an additional layer of complexity in archaeological objects which is not easily accounted for in the conventional geological approach. Using the techniques described in Chapter 6, we can distinguish between objects where the lead is low and the lead isotopes are likely to reflect the source of the copper, and those objects containing more lead, probably deliberately added, where it is the source of the lead that is being identified by the isotopic data. Since such diagrams also show *mixing lines* between different isotopic sources of lead, we can begin to see patterns where the same source of copper is mixed with lead of two or more isotopic values, potentially reflecting two or more different sources of lead. Equally, we can see the reverse—the same lead being mixed with two different sources of copper. These examples simply serve to show the complexity of the possible metal flow patterns that might occur within and between different societies, but also that the methods described here can begin to unravel this complexity.

We can now re-visit the idea of an *assemblage,* introduced above as being the totality of the objects from a particular place and time for which we have chemical and/or isotopic data. We use such an assemblage as the best possible proxy for characterizing the metal available at a particular place and time. We must, of course, always remember that our assemblage is, at best, a biased sample drawn from a biased sample of a biased sample of an unknown and unknowable parent population! The three sources of bias referred to here are i) the bias introduced by the original choice of objects to be deposited into archaeologically accessible contexts, ii) the bias of archaeological recovery in terms of the contexts selected for excavation, and iii) the bias in selecting excavated objects to analyse chemically and isotopically. As explained in Chapter 3, it requires a strong focus on typology and archaeological context when interpreting the data from such analysed objects to minimise the effect of these biases.

Whilst this is one way of using the term assemblage, we can sometimes be more specific, which might also be helpful in countering some of the biases discussed above. It is essentially a scalable parameter which needs to be specifically defined for each question being asked. For example, it could be all the metal objects from a single tomb, which then allows us to compare the characteristics of the metal in this tomb with those from other tombs, or with the general pattern of metal in circulation. Within the excavation of a single site, it could be all the metal objects from a particular phase of occupation, which would allow us to look at changes over time by comparing the metal assemblage between phases at that particular site. Scaling up, we can equally define an assemblage as being all the metal from a particular cultural group, which allows us to compare between groups—thereby addressing questions of the degree of interaction between adjacent cultures, or the degree of continuity between successive cultures. On the other hand, we may choose to classify all the objects of a particular type as an assemblage, irrespective of where they were found, thus allowing questions of the relationship between typology, function and metal use to be considered (such as 'were personal ornaments made from the same copper, and alloyed in the same way, as weapons'?). In each of these cases, the assemblage is selected specifically to represent the class of objects necessary to answer the question being asked. When combined with *ubiquity analysis* (the percentage of a particular assemblage made up of a particular type of copper or alloy) and *profile analysis* (the distribution of a particular element in all of the objects in the assemblage) we can use spatial and temporal mapping to follow these subtle chemical shifts caused by human interventions through space and time.

Taken together these tools offer an integrated methodology which combines: i) a model for the chemical changes in copper-alloys caused by human and technological processes, with ii) a re-definition of the terminology for alloy composition, which does not implicitly assume deliberate alloy design, and iii) a new way of interpreting lead isotope data that is more sensitive to anthropogenic mixing. One strong feature of this system is that it is both *scalable* and *universal*. It is *scalable* in the sense that it can be applied to assemblages representing the contents of a single grave or hoard, up to a particular type of object which is distributed across all of Eurasia. It is *universal* in the sense that the basic methodology can be applied anywhere, and used to compare assemblages from widely separated places and times. That is not to say that it can be applied anywhere in a mechanical fashion, with guaranteed outcomes—although the *processes* we have developed are universally applicable, the interpretation of the observed changes will be radically different, depending if one is dealing with a set of relatively small-scale loosely organized societies, such as those found on the Steppe, or a highly organized and centralized state such as Dynastic China or the Roman Empire. Nor can it be assumed that by simply applying the prescribed methodology to any archaeological situation, all questions will be answered! It does mean, however, that data from, say, Eastern Europe can be directly compared with that from southern Siberia. This avoids the limitation seen when classifications taken from earlier studies are compared, since most of these classifications are derived from internally defined parameters, making them specific to that dataset. Thus the outcomes of the SAM programme cannot be directly compared with those of Chernykh, whereas using our methodology the results can be directly compared across all of Eurasia. Our assertion is that the methodological tools described here can be used as a starting point for any archaeological interpretation, but specific questions might require a different set of subsequent approaches.

Sample bias

Above we raised the issue of the potential bias between the objects for which we have chemical analyses, and the totality of the metal produced in a particular region. This is a serious issue, and one that is rarely discussed in traditional approaches to archaeometallurgy, particularly in provenance studies. One conventional aspect of sample bias is to consider how well the average analysis of the assemblage (the sample) represents that of the (unknown) parent population (the flow). We can of course calculate the average composition of the objects in the assemblage available to us, and produce a mean and standard deviation for each element (e.g., Cu = 64.5 ± 1.2 %, etc.), providing that we think the distribution of each element is approximately normal within the assemblage. Sampling theory, however, tells us that although the average is the best available estimate for the mean of the population, the standard deviation of the *sample* (the analysed assemblage) is not the standard deviation of the *population* (i.e., the stock of metal), and also that the calculated standard deviation for the assemblage is invariably smaller than that of the parent population (Miller and Miller 1984, 41–44). The latter can be calculated from that of the sample, provided we know the sizes of the sample and the population, as illustrated in **Figure 4**. We know the size of the sample, but that of the population is unknown, and is likely to be much greater than that of the sample. If the sample size is large compared to the assumed size of the parent population (e.g., if we have analysed most of the objects in a tomb), then the difference will be minimal, but for small samples, where we might only have analysed 200 objects from an area which is likely to have produced millions, then the difference will be very large. Most archaeometallurgical studies, however, do not take this into account, and simply take the parameters of the sample to be those of the parent population, and then use these data to perform further numerical calculations.

A more significant issue in the context of FLAME is the potential bias arising from the typological mismatch between the sample and the parent population. As shown in **Figure 5**, if the hypothesised parent population contains a number of different typologies, but the proportions in each segment are unknown, then we have to assume that the totality of *known* objects (i.e., those which have been archaeologically recovered) represents faithfully the divisions in the unknown parent population. In this hypothetical example the recovered population consists of 25% axes, 35% daggers, 10% swords and 30% pins by number. If we assume that each of these categories has a different chemical composition, then the analysed sample will be biased if it does not contain the same proportion of object types. This becomes an important consideration when looking at regional alloying practices in Chapter 5, and is discussed further there.

In the light of the difficulties of predicting the chemical properties of an unknown parent population from an inevitably biased sample of those objects which have been excavated and chemically analysed, we might be tempted to give up, although all archaeological research in one way or another has to learn to deal with such challenges. In fact, one of the characteristic features of archaeology is that it has to come to terms with data that are far from ideal in the statistical sense. It generally does this by recognizing the limitations of the data (often, however, implicitly) and devising ways of overcoming them. Mathematically speaking, a good start is often provided by switching from parametric to non-parametric statistics—i.e., away from using means and standard deviations to characterize the data, and using medians, interquartile ranges and order statistics instead. The methods we describe here are essentially non-parametric and do not rely on using descriptions based on means and standard deviations, with the concomitant assumptions of normality. They are therefore inherently better suited to dealing with the sort of data that we routinely encounter. They are also reasonably robust with respect to errors in the actual measurements, as described in the next chapter. In short, we argue that the approach described here is not only conceptually more useful when considering the role of metal within human society, but, given the nature of the data, is also mathematically more appropriate.

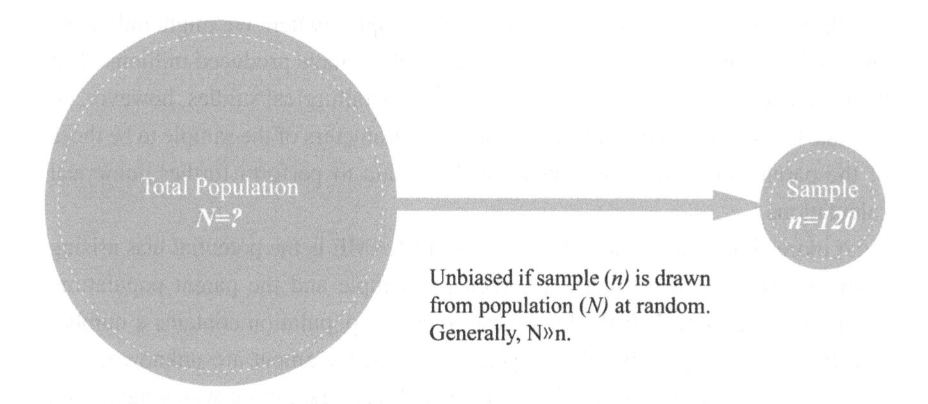

Figure 4:
Relationship between the sample (assemblage of analysed objects) and the parent population (the stock of metal) if the sample is unbiased.

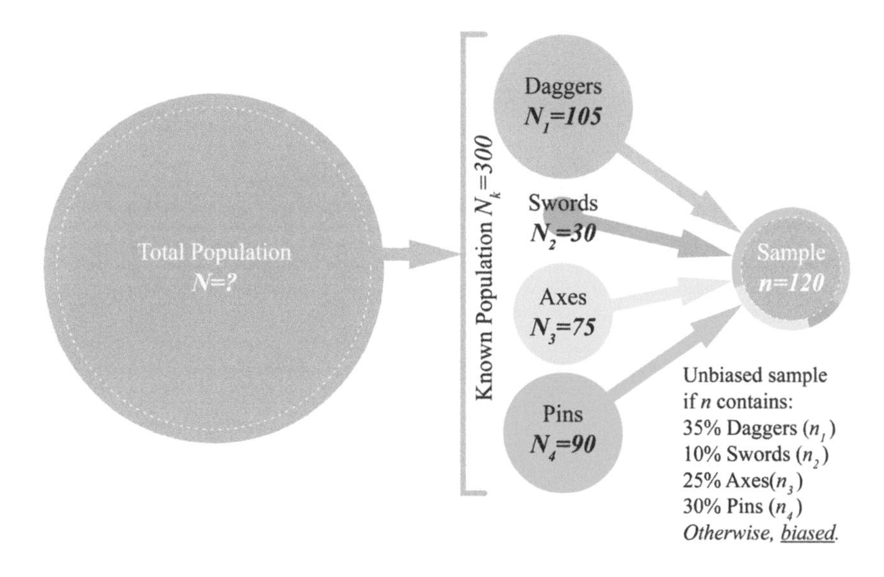

Figure 5:
Relationship between a more complex parent population (stock of metal) and the assemblage.

Chapter 3

Legacy Datasets and Chemical Data Quality

As explained in the previous chapter, the 'Oxford system' focusses not on the analyses of individual objects, but on the properties of specific assemblages, the definition of which depends on the nature of the question being asked. It is inherent in the methodology that the larger the dataset the more reliable will be the inferences derived from it, subject of course to the requirement that the assemblage (sample) is representative of the objects required to address the question under consideration (population). It is this quest for the largest possible dataset that has drawn us into considering the potential for the use of *legacy datasets*—chemical data compiled from published sources, some of which may be old, and perhaps using now obsolete methods of analysis. The obvious alternative would be to restrict the analysis to only high quality modern data (although few datasets would actually fully meet the highest possible standards of '*analytical hygiene*', as set out below). However, this would probably reduce the volume of data available for Bronze Age Eurasia from more than 100,000 analyses to fewer than 10,000, and would also mean that vast areas of the continent would have no representative data at all. It would of course be ideal to initiate a new programme of chemical and isotopic analysis of Bronze Age metalwork to the highest possible modern standards, but the cost, time required, and difficulties associated with obtaining sampling permission means that this is unlikely to happen in the near future, if ever. We are therefore presented with a dilemma—either find a way of using a heterogeneous compilation of legacy data which gives the largest possible dataset, or only use high quality data but with fewer numbers and smaller geographical coverage. For a project which attempts to cover all of Eurasia, we have chosen the former, but we can easily see that more specific projects might beneficially be able to use the latter.

Evaluating 'legacy data'

The use of legacy chemical data compiled from published sources gives rise to the obvious question of consistency between datasets. **Table 1** lists the most common methods of chemical analysis used on archaeological copper alloy objects, and highlights some of their strengths and weaknesses. For a more complete discussion of the various methods of analysis see Pollard *et al.* (2007; 2017). Apart from the fact that not all analysts have reported data on what would now be regarded as the minimum set of elements (for copper alloys, the measurement of Cu, Pb, Sn, Zn, Fe, As, Sb, Ag and Ni), it is also well-known that some analytical methods have systematic problems with certain elements. Moreover, virtually none of the published literature contains sufficient information on the primary or secondary standards used, nor the estimated levels of detection, precision, or accuracy. How, then, can such datasets be combined?

One of the first problems to be overcome is to ensure that the data we are using contain valid estimates for all of the elements we think are reported. In the oldest literature, where measurements were done by gravimetric analysis, if an element was not measured it was often just left out of the results list. Hence, many of the earliest analyses of archaeological bronzes (from 1777 – *c.* 1830) simply report copper and tin, or sometimes only tin, with copper calculated by difference from 100% (Pollard 2016). Later in the history of gravimetric analysis, it became common to measure more than five elements (often Cu, Sn, Pb, Zn and Fe, perhaps with Ni and some other trace elements), although many were recorded as 'tr.' or '-'. There is no uniformly accepted convention for using such codes to differentiate between all the possible variations of 'not looked for', 'not determined', 'not detected', 'trace', or just 'absent' (i.e., below the level of detection of the instrument). Consequently, it is not always easy to interpret the meaning of symbols such as '-' in some datasets. Does it mean 'absent' or 'not detected' (i.e., below an invariably unspecified level of detection), or 'not looked for/not determined' (i.e., no attempt has been made to measure it)? If it is the former, then we might be able to make an educated guess about the likely limit of detection for a particular element by that particular analyst. For example, if we have sufficient data from that analyst, then we might look for the lowest recorded value of a particular element, and take that to be the limit of detection. We can then record the element reported as '-' as being below that value. If it was not looked for, then we have to mark it as 'unknown', unless we can be reasonably certain that all the other significant elements have been measured, and that the data have not been normalised, in which case we might use the difference between the analytical total (the sum of all the elements measured) and 100% as an estimate for that element. This is, at best, a very uncertain process, because all the measurement errors for each element will be accumulated into any such estimate by difference, meaning that the uncertainty could be much larger than the estimate. Moreover, it assumes that the only non-measured element is the one of interest. We would certainly not recommend using such a procedure for any trace elements, and only *in extremis* for the major alloying elements. Most commonly this approach is used where copper itself has not been directly measured, and is estimated by difference (as in the SAM data). It is probably better to use this estimate by difference for copper and retain the dataset rather than to discard the analyses completely, but it should always be recorded that the copper was not directly measured, and is likely to be less accurate than if it had been.

Method	Comments	Strengths	Weakness
Gravimetry (Wet Chemistry)	Precipitation and weighing of specific elements from solution	Large sample so gives bulk analysis	Early work (pre-1900?) is generally unreliable for trace elements. Elements not specifically looked for are missed.
Optical emission spectrometry (OES)	Spark or arc emission and photographic recording in the early years (1920–?1960)	Records emission lines of all elements present	Reproducibility of photographic plate is poor. Emission lines of major elements saturate at high concentrations and become non-linear if not done by separate exposure
Atomic absorption spectroscopy (AAS)	Solution input, analysis done sequentially element by element	Good sensitivity for many elements	Time consuming. Elements not looked for are unrecorded
Electron microprobe (EM, microprobe, EPMA)	Solid sample mounted in microscope. Wavelength dispersive detection of X-rays	Spatially resolved analyses. Better sensitivity then SEM	Surface sensitive. Spectral overlap between certain elements. Tendency for elements not looked for to be missed.
Scanning electron microscopy (SEM)	Solid sample mounted in microscope. Energy dispersive detection of X-rays	Spectrum records all elements present simulataneously. Spatially resolved analyses.	Surface sensitive. Significant spectral overlap between certain elements. Poorer sensitivity than EM.
Wavelength-dispersive X-ray fluorescence (WD-XRF)	Solid sample or pressed pellet. Wavelength dispersive detection of X-rays	Better sensitivity then ED-XRF	Surface sensitive. Spectral overlap between certain elements. Tendency for elements not looked for to be missed.
Energy-dispersive X-ray fluorescence (ED-XRF)	Original object, solid sample or pressed pellet. Energy dispersive detection of X-rays.	Records all elements simulataneously. Rapid. Can be 'non-destructive'	Surface sensitive. Significant spectral overlap. Poorer sensitivity then WD-XRF. Quality depends on degree of surface preparation on whole objects.

Portable X-ray fluorescence (pXRF)	Orignal object. Energy dispersive detection of X-rays.	Portable. Rapid.	Surface sensitive. Significant spectral overlap. Poorer sensitivity then WD-XRF. Quality depends on degree of surface preparation on whole objects. See discussion in text.
Neutron activation analysis (NAA)	Needs nuclear reactor to activate samples. Gamma ray detector.	Extremely sensitive	Incapable of measuring certain elements esp. lead. Leaves samples radioactive for many years.
Inductively-coupled plasma emission spectroscopy (ICP-OES)	Solution input, emission line detection via CCD	Highly sensitive	Although emission spectrum contains lines for all elements present, recording usually only for specified elements. Emission lines overlap for certain elements
Inductively-coupled plasma mass spectroscopy (ICP-MS)	Solution input, mass spectrometric detection using either quadrupole or multiple ms	Extremely sensitive, especially with multiple mass spectrometers. Capable of isotopic ratio analysis.	Isobaric peak overlap. Quadrupole records all elements present, mulitple ms usually only specified isotopes.
Laser Ablation-Inductively-coupled plasma mass spectroscopy (LA-ICP-MS)	Solid sample ablated by laser, mass spectrometric detection using either quadrupole or multiple ms	Extremely sensitive, especially with multiple mass spectrometers. Capable of isotopic ratio analysis. Spatially resolved analyses possible.	Isobaric peak overlap. Quadrupole records all elements present, mulitple ms usually only specified isotopes. Fractionation during ablation process. Difficulties of calibration.
Proton-induced X-ray emission (PIXE)	XRF using proton stimulation rather than X-rays	Sensitive with good spatial resolution	Requires proton accelerator. Spectral overlap between certain elements.

Table 1:
Summary of the strengths and weaknesses of various analytical methodologies commonly used on archaeological metals.

Similar considerations apply to the interpretation of entries such as 'tr' ('trace'), or semi-quantitative results such as '+', '++', '+++', etc. The process of interpretation of such hieroglyphs requires considerable knowledge of the historical and analytical context of the data, such as what instrument was used, and to some extent what standards might have been available. It is sometimes possible to reconstruct what 'tr' might mean in a particular context by following the procedure described above, i.e., finding the lowest recorded value for a particular element in that context, and assigning 'tr' to that value, or just below it. The extensive use of semi-quantitative data (such as +, ++, +++, etc.), as was done by Pittioni's group in Vienna (see Chapter 1), is considerably more problematic. Apart from Pittioni's work, where it was used for all data, it is mainly a feature of older work such as that produced before the 1970s by optical emission spectroscopy using photographic plate recording. Here the instrument settings were usually optimised for trace elements, since in the early days of optical emission it was seen as a way of providing better trace element data than could be obtained by conventional gravimetric analysis. Unless separate readings with different settings were taken for major and trace elements, as was done by Chernykh, the major elements tended to 'burn out' (saturate) the plate, resulting in nonlinearity in the recorded intensity. This is the reason why some early instrumental measurements of tin concentrations tend to saturate at around 12–15%, and the result is simply recorded as '>12%'. Some attempts, such as those by Ottaway (1982) described below, have been made to convert these semi-quantitative symbolic recording systems into a quantitative scale, but they cannot be regarded as accurate or satisfactory.

One further decision that often has to be taken when considering legacy data is how to use data such as 'Sb<0.05%' (and, more rarely, 'Sn>15%', as discussed above) in any quantitative analyses. Most database software will not handle any non-numeric characters during calculations, so they either have to be omitted (and usually the entire analysis removed, to stop the calculation crashing), or replaced with a finite quantity. A convention that is commonly used with such data is to systematically replace any minimum estimates (denoted by '<') with half the given value. Thus '<0.05' would be replaced by '0.025'. Although clearly not at all accurate, this procedure does at least allow the essence of the data to be retained. Values recorded as 'minimum maximum estimates' (e.g., Sn '>15%') are more difficult to replace, since the realistic upper bound is generally not known (in principle of course it is 100% minus the sum of the quantified elements, but in practice it is difficult to estimate if the data have already been normalised, or the copper already estimated by difference). Often the only recourse is to take an average of the value for that element in similar objects where a definitive value has been recorded and use that, but this is clearly only satisfactory in that it allows the data to be retained in the calculation, and is highly unlikely to be accurate.

As described in the following two chapters, our methodology for both trace elements and alloying elements has, as its initial stage, a presence/absence classification. This is discussed in detail later, but normally we would use 0.1% as the cut-off value for trace elements (As, Sb, Ag and Ni), and 1% for the alloying elements (tin, lead and zinc: see Chapter 4 for a discussion about arsenic being considered as a deliberate alloying element). Of relevance to the above discussion, we note that this simple preliminary binary characterization stage reduces some of the uncertainty introduced by the approximations which have to be made when attempting to interpret the undefined lower values in the data. Thus, if we estimate the lower limit of detection of nickel in a particular analysis to be 0.06% (by looking for the lowest reported definite value), then anything labelled 'tr' in that analysis is recorded as <0.06%. If we then remove the non-numeric symbol and replace this value by 0.03% to allow calculations to be carried out, then it is clear that allocating this sample to the category 'Ni = absent' on the basis of a 0.1% cut-off is likely to be correct, despite the uncertainties introduced by the successive approximations. Likewise, any tin value recorded as '>15%' can still safely be allocated to the alloy type 'Sn>1%', irrespective of how we estimate the true tin value. Subsequent stages of the analysis of copper groups or alloying types are certainly susceptible to bias if there are a large number of such estimated values in the assemblage. Nevertheless, they are likely to be less affected by these issues than more traditional numerical techniques (such as principal components analysis, or cluster analysis), because we use techniques which are relatively insensitive to individual values, such as ubiquity analysis, where we calculate the percentage of the assemblage which is of a particular copper group or alloy type. When doing profile analysis, we use histograms and cumulative frequency distribution functions (or create kernel density estimates of the distribution), which are also relatively insensitive to specific individual values, when compared to the traditional approaches used in clustering.

As discussed in more detail in Chapter 7, our practice is to record the data exactly as it is published in the base layer of the database, and not to allow any modifications to this layer. We then generate a second layer to be used in calculations, where any obvious errors are corrected in the data (typos, switches of columns, etc.), and also where any changes to remove non-numeric characters can be made, according to a specified protocol such as that outlined above. Essentially, the interpretative procedures we are using are, initially at least, non-parametric approaches rather than the more usual parametric methods (which include clustering algorithms), which are always going to be more tolerant of imperfect data. Thus our methodology is more robust, which allows us to use more of the legacy data than would otherwise be the case. Clearly the problems are not completely eliminated, however, and care must be taken when the values for the trace elements or alloys fall close to our cut-off values (a calculated minimum value of 0.1% or greater for a trace element would clearly create difficulties in the allocation to a copper group). Methods for dealing with such situations are discussed in the next chapter.

An early misinterpretation of our methodological approach was that, by using the methods described in the next two Chapters, we are advocating a return to the system of semi-quantitative analyses, as carried out particularly by Pittioni and his colleagues in Europe in the 1950s and 1960s. Their system was chosen partly because they felt that absolute values were unnecessary, relying instead on patterns of ratios, and necessitated by the fact that in the earliest days of emission spectrometry it was difficult to quantify the intensity of the emission lines, so they reported all of their data on all elements as one of six ranked symbols (-, ?, Sp, +, ++ and +++). We have never advocated a return to such a system. As is made clear in the following chapters, our methodology *starts* with a preliminary presence/absence classification of the trace elements in order to determine the major patterning in the assemblages, but continues with the use of elemental profiles, which require a fully quantitative analysis. Likewise, for alloying elements, we start with a presence/absence classification, but again follow this up with the use of elemental profiles. Thus our system is predicated on the availability of fully quantitative data. The fact that all of Pittioni and his colleagues' data is presented only in a semi-quantitative way is extremely unfortunate from our point of view, because they measured a wide range of elements on around 6000 objects, but their method of reporting makes it impossible to integrate these data with any other data, and they can therefore only be of the most generic value.

The difficulties associated with using data from several sources, especially if some of them are generated by obsolete analytical procedures, cannot of course be underestimated or ignored. As discussed above, however, we argue that the benefit of using as much data as possible counterbalances to some degree the loss of data quality. The most likely exception to this is likely to be a vulnerability to variations in limits of detection around our arbitrary cut-off values, but, as described in the following chapter, systematically varying the cut-off around the value of 0.1% allows us to check that the results are not unstable with respect to the choice of cut-off. Additionally, by retaining as much data as possible, we can ensure that our interpretations are based on assemblages created from a large number of analyses. Thus we might expect that errors created by conflating data from several sources might partially cancel out—if, for example, one method tends to underestimate a particular trace element, whereas others correctly estimate or even overestimate it, then given enough data from as many different analytical sources as possible, the overall picture should converge on an 'average' set of values. Of course, this is not an infallible defence against bias, since for some sites and periods the data may be from a single analyst and by a single method, but it is some protection against rogue data.

Further evidence for the validity of our approach is provided by the fact that it allows us to detect and isolate discrepant datasets, simply by checking for subsets which do not fit the general pattern (often by a particular analyst) within our large dataset. Thus we were able to see that von Bibra's data published in 1869 on Roman and Byzantine coinage was unreliable with respect to nickel—simply by comparing his data with the large volume of more recent data on the same types of coins. He reported the presence of nickel in most of the coinage, whereas other analysts found essentially none. Interestingly, after we had deleted von Bibra's trace element data from the analysis, we discovered a comment by Caley in his book on Greek coinage which said that von Bibra's gravimetric procedure was unreliable for the trace elements, especially nickel (Caley 1939, 154). In this case, because in gravimetric analysis each element is determined independently, and von Bibra did not normalize his analyses to 100%, we were able to retain his major element data in the analysis of Roman and Byzantine coinage, but removed all of his trace element data from our subsequent calculations.

To summarize, we believe that the difficulties involved in using old data are more than offset by the advantage of being able to use large datasets, providing an appropriate method of interpreting the data is used. This allows the inclusion of data from objects for which it would now be impossible to obtain better quality data, given the modern tendency in most museums to restrict analysis to 'non-destructive' methods. The system described here has the advantage that it can be applied without re-sampling, and allows us to use the archaeometallurgical equivalent of a 'Big Data' approach—not on the scale that any data scientist would recognize as "Big Data", but certainly on a scale hitherto unavailable to archaeological materials scientists.

'Round robins' for the analysis of archaeological bronzes

It is well-known in analytical chemistry that if the same object is analysed by different laboratories it is not necessarily guaranteed that they will provide identical results. This could be due to the use of different types of instruments, or differences in sampling and calibration procedures, or simply due to the experience of the analyst. The question of inter-laboratory consistency has been repeatedly investigated in the western archaeometallurgical literature. The first international evaluation of data quality between different laboratories analysing ancient bronzes was that reported by Chase (1974). Three samples (one Chinese Shang Gu vessel, one Luristan bronze spear point and one modern standard) were circulated amongst 21 laboratories and analysed for as many as 48 elements. It is unfortunate that not all the participating laboratories reported the information required by Chase, so that the final publication merely reported on the quantification of copper. This revealed a clear systematic difference in the reported copper values, showing that several laboratories tended to overestimate the percentage of copper, but without the full dataset it is not possible to evaluate the overall quality of the data.

Another important and fully documented inter-laboratory comparison on bronzes was that reported by Northover and Rychner (1998). The same set of seven archaeological objects (4 sickle blades, a winged axe, a knife and a bracelet) was analysed by 22 laboratories using 24 different analytical methods (1 wet chemistry, 3 OES, 4 ICP-OES, 1 direct current plasma spectrometry, 4 AAS, 4 ED-XRF, 4 EMPA, 1 PIXE and 2 NAA). The circulated objects contained trace elements from 0.1% to 0.5%, which allowed the comparison of precision and accuracy when the results approach the detection limit of the various instruments. They carefully examined the performance of each instrument on a comprehensive list of elements and concluded that:

> "*we have successfully demonstrated that modern analytical techniques are capable of producing accurate, reproducible data that can, with thorough standardization, be used interchangeably with other data and behave similarly in clustering and classification*" (Northover and Rychner 1998: 31).

The most recent set of comparisons are those reported in a series of papers published by Heginbotham *et al.* (2011; 2015; in press) and Heginbotham and Solé (2017). The aim of this series is different from the previous more general comparisons between different instrumental techniques, in that it is focussed specifically on improving and harmonising the use of X-ray fluorescence (XRF) on museum copper alloy samples with minimal sample preparation. Heginbotham *et al.* (2011) compared data produced by XRF from different laboratories using different calibration methods and standards as a baseline for further development. Fourteen museum institutions and a total of 19 different XRF systems were involved in this study. A set of 12 samples (four cuttings from standard reference materials, six pieces of historic metal and two small ingots prepared by the first author) was analysed in a round-robin process by all of the participating laboratories. Participants reported instrumental details, analytical protocols and parameters alongside the analytical results. The overall inter-laboratory reproducibility was found to be low, suggesting that data from the different laboratories could not be comfortably aggregated into a single comprehensive database. Most of the best-performing laboratories used a fundamental parameters (FP) calibration procedure to convert the original X-ray counts into elemental percentages, plus calibration using standards with known chemical concentrations. Heginbotham *et al.* (2015) described the design and production of the CHARM (Cultural Heritage Alloy Reference Materials) reference standards. The 'core' set consists of 12 standards with certified values for Cu, Zn, Sn, Pb, Fe, Ni, As, Ag and Sb, plus S, Cr, Co, Se, Cd, Au, Bi. Two 'extension' sets are also available, one for higher As (up to 3.66%, with two standards) and one for cupro-nickels (up to 30% Ni, with three standards). Heginbotham and Solé (2017) then propose a comprehensive protocol for energy dispersive XRF (ED-XRF) quantification of museum copper alloy objects using free open-source fundamental parameters software for spectral analysis (PyMca) combined with the CHARM reference set.

Data inter-comparisons

As well as these 'round robins', where several laboratories are invited to analyse the same materials, a number of targeted data inter-comparisons have been carried out, with the specific objective of assessing whether data produced by one method can be directly compared with those produced by another. For example, Hughes (1979) reanalysed using AAS some Middle and Late Bronze Age objects previously measured by OES, reported in Brown and Blin-Stoyle (1959), and found comparability for the elements Sb, Ag, Ni, Bi, Fe, As. However, the disagreement in the concentration of lead by these two techniques was shown to be large. Hughes concluded that OES is likely to underestimate lead levels in heavily leaded bronzes, possibly due to the self-absorption of lead emission lines in the vaporized sample. Normally this would be expected to be revealed by an abnormally low analytical total in the OES data, but if the data for a single sample are normalized to 100%, it will be obscured and give rise not only to a low lead concentration, but also to an over-estimation of other elements. In fact, in the data presented by Hughes (1979, table 1), the samples with low lead by OES compared to the AAS results do not have low analytical totals, but nor do they seem to have been normalized. This suggests that the discrepancy could be more due to inhomogeneity in the objects with higher lead levels rather than a systematic difference between OES and AAS (see below).

In a major project on the prehistoric use of copper in Bulgaria, Pernicka *et al.* (1997) took the opportunity to compare OES analyses conducted by Chernykh (1978) in Moscow with neutron activation analysis (NAA) data produced in Heidelberg on the same objects. **Figure 1** shows a comparison of the two sets of results, redrawn from the original publication by Pernicka *et al.* (1997, 89). The majority of NAA results agreed with the prior OES analyses within a factor of three, indicated by the shaded band. However, drawing the conservatively estimated minimum detectable level (MDL) lines for NAA (ca. 10 ppm) and OES (ca. 500 ppm) on each diagram excludes a large number of points showing poor agreement, which reinforces the observation of general agreement made by the original authors. Because NAA is insensitive to lead, the Heidelberg lead analyses were conducted using isotope dilution mass spectrometry. As seen by Hughes (1979), there was a larger disagreement in lead concentration results between the earlier and later analyses. Although the overall impression from this comparison is good, we should note that agreement to within a factor of three is not a very high level of consistency.

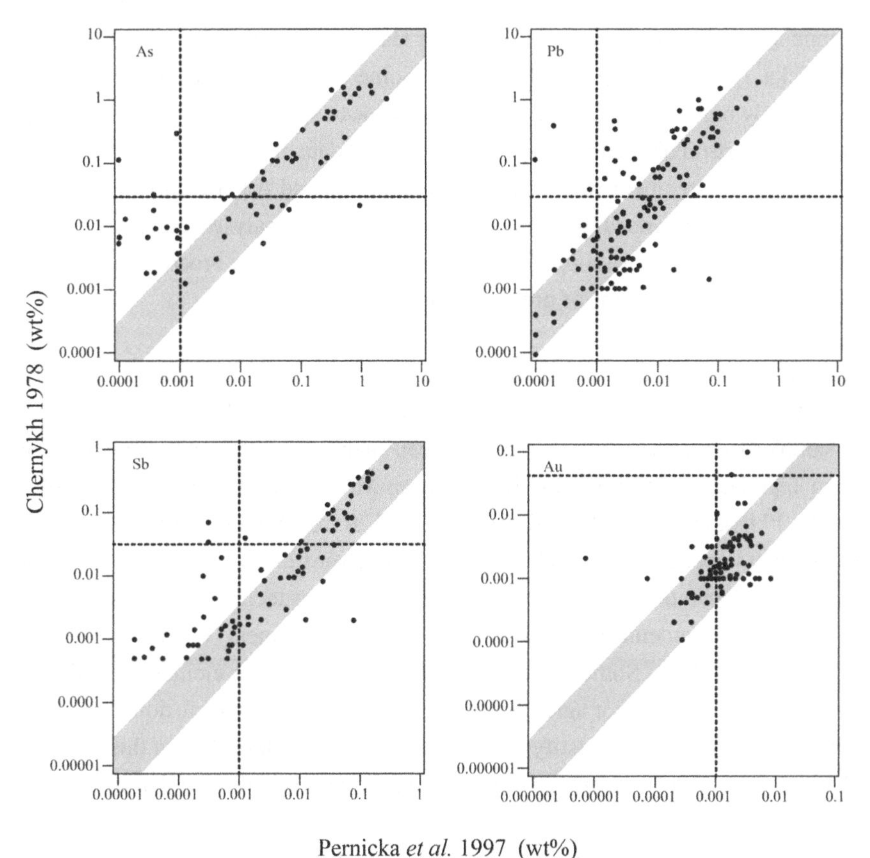

Figure 1:
Comparison of chemical composition from the same Bulgarian artefacts analyzed by OES (Chernykh 1978, Moscow: y axis) and NAA (Pernicka ***et al.*** 1997, Germany: x axis)[1]. The shaded band indicates the respective concentrations in agreement within a factor of three. Redrawn from Pernicka ***et al.*** (1997, fig. 15). Dashed lines are added to indicate the likely MDL for each element by each method (OES from Pollard ***et al.*** 2018, Table 1; NAA from Hancock ***et al.*** 1991).

[1] Lead was measured with isotope dilution mass spectrometry by the German team.

Perhaps the most significant comparison to be carried out, at least as far as European archaeology is concerned, is that of Pernicka and his group at the Max Planck Institute in Heidelberg between 1986 and 1997. Under the title "Stuttgarter Metallanalysen-Project" (SMAP) they inherited the data and samples from the SAM project (see Chapter 2). Because they wished to add new data to the SAM database, and also to re-investigate the old data, a key issue was evaluating the intercomparability of old and new data. A comparison of the SAM data (by OES) with new analyses in Heidelberg by NAA (for Ni, Ag, As, Sb and Sn) and atomic absorption spectroscopy (AAS) for lead and Bi yielded good correlations (Pernicka 1986, fig. 2, p. 27: **Figure 2**). Some anomalies were noted, such as erroneously high values for antimony by OES due to spectral interference from Fe, and a comparison of electron microprobe (EMPA) data on a smaller set of samples showed poor agreement with NAA data for Sn, Ag and Au (Pernicka 1986, 28). However, as summarised by Krause (2003, 18–22), the analyses undertaken using optical emission spectroscopy (OES) in the SAM projects are broadly comparable with modern data in terms of precision and detection limits.

Another important comparative project was carried out by Cuénod (2013, 139-147) on data from the Bronze-Iron Age transition in Iran. She collated a large number of samples from specific sites which had been analysed by different techniques (56 from Tepe Hissar (OES and PIXE), 151 from Mesopotamia (XRF and AAS plus NAA), 17 from Selme (AAS and ICP-OES) and 70 from Susa (UV-OES and spark-source mass spectrometry)). Through straightforward plotting of the analyses of individual elements from specific sites by these various techniques, she suggested that in spite of undeniable differences, a 'general agreement' could be observed (Cuénod 2013, 146). Some of her comparisons for particular elements are shown in **Figure 3**, showing poor agreement for lead (Fig. 3d). Her conclusion from these comparisons was used to justify the use of the approach described in this volume: although there was 'general agreement', the differences were such that it:

> "*dissuaded us from using automated methods of analysis, such as clustering and principal components analysis, as they are likely to pick up technique-related trends rather than archaeological ones*" (Cuénod 2013, 146).

This is a clear statement of one of key drivers for developing an interpretative methodology which can tolerate differences between analytical methods.

Apart from these major projects, a variety of smaller scale comparative studies have been conducted by many individual laboratories (e.g., Gilmore and Ottaway 1980, Ottaway 1982, Carter *et al.* 1983, Lutz and Pernicka 1996, Northover 1999,

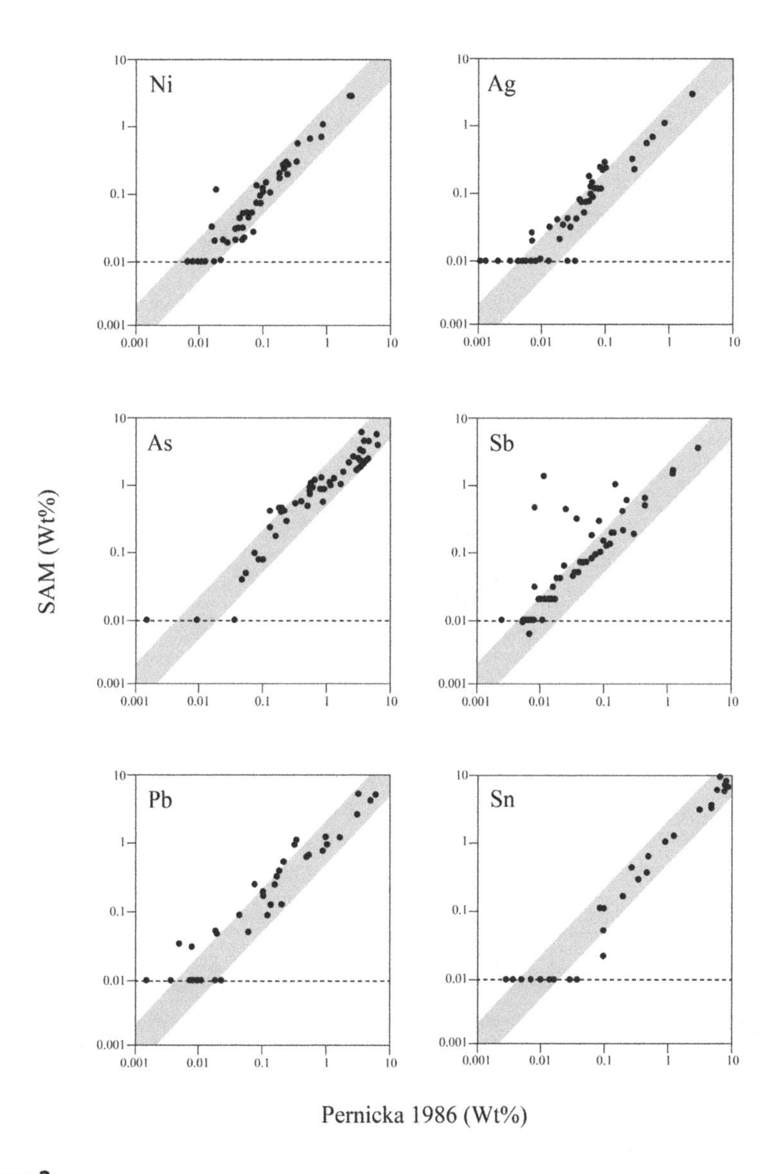

Figure 2:
Comparison of SAM (OES) and Heidelberg (NAA or AAS) data on the same objects (redrawn from Pernicka 1986, fig. 2, p. 27).

Cooper *et al.* 2008, Willett and Sayre 2000, Young *et al.* 2010). However, it is still necessary to bear in mind that any round-robin experiment or analytical intercomparison is merely a snapshot of the data quality of these laboratories at the time of analysis, rather than a routine monitor of data quality. This difference is discussed further below.

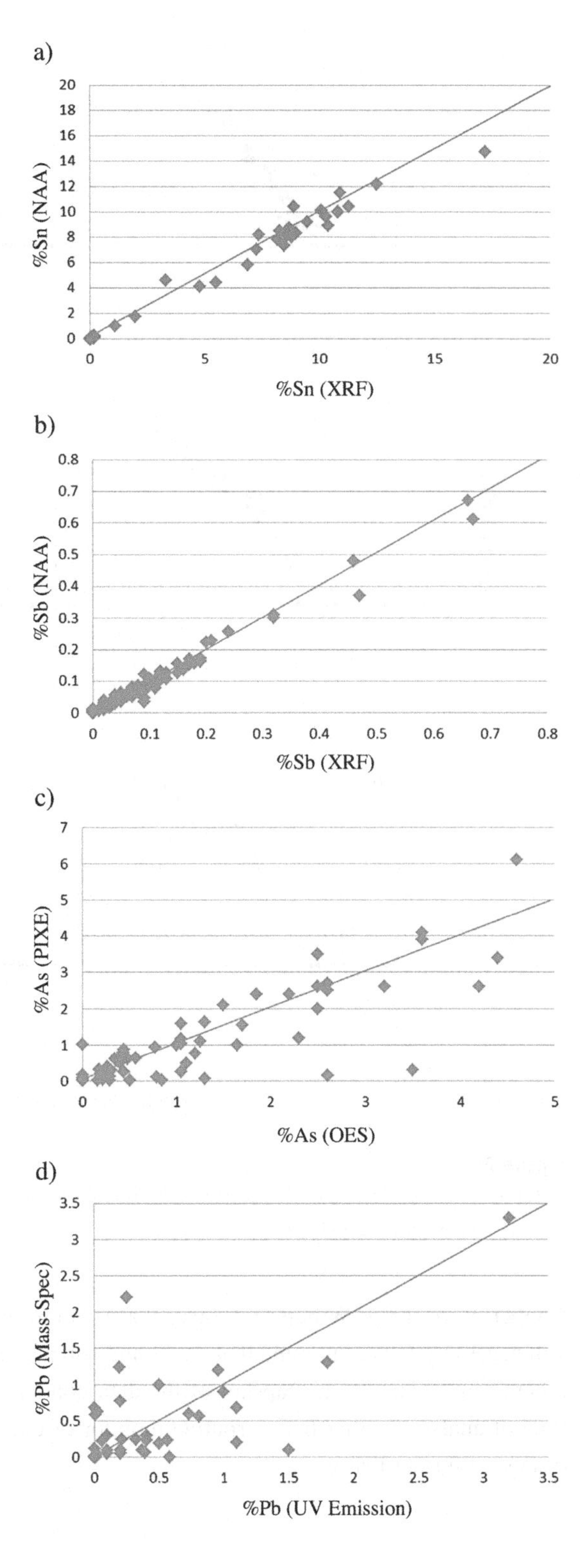

Figure 3:
Comparisons of particular elements from specific sites in Iran and Mesopotamia using different analytical techniques (Cuénod 2013, 139–147). a) tin in samples from Mesopotamia, XRF vs. NAA; b) antimony in samples from Mesopotamia, XRF vs. NAA; c) arsenic in samples from Tepe Hissar, Iran, OES vs. PIXE; d) lead in samples from Susa, Iran, UV emission spectrography vs. mass spectrometry.

Portable XRF (pXRF) data on archaeological bronzes

Many companies now manufacture lightweight hand-held X-ray fluorescence machines with 'point and shoot' capabilities (Shugar and Mass 2013), which were initially developed to be used by non-specialist analysts for specific industrial uses. The 'rise and rise' of the applications of such instruments in archaeology has caused much discussion and controversy (Shackley 2011, 1–6). Portable XRF machines have many advantages, the most obvious being that they allow the analyst to go to the object rather than vice-versa, and they can produce completely non-destructive instant analyses. These attributes make pXRF particularly attractive for curators of museums and collections, which a recent survey of the literature has shown to be the most common locus of application (Frahm and Doonan 2013).

Certain areas of research, such as obsidian studies, have enthusiastically adopted pXRF to great effect—a review published in 2013 by Speakman and Shackley listed 78 papers using pXRF on obsidian. Metals in principle are good for analysis by pXRF, since most of the elements of interest are in the mid- to heavy part of the periodic table, and therefore the characteristic X-rays are not unduly absorbed by passing through an air path. The critical issue for pXRF analysis of archaeological metals is whether the surface is cleaned or not before analysis. In principle, there is no disadvantage to using pXRF on prepared metal surfaces, although it should not be expected to be as sensitive as similar data collected by larger XRF machines or SEM because of the lower intensity of the primary beam, the larger irradiated area, and the absorption of both the primary and secondary beams by air. However, several studies, including our own, have shown that pXRF on *unprepared* metal surfaces can give grossly misleading results—especially if the surfaces are visibly corroded.

Hsu (2016) analysed by pXRF 193 metal objects from the Bronze and early Iron Age of the eastern Eurasian steppe in the British Museum, of which 75 had previously been analysed by OES in the late 1950s (Pollard *et al.* 2018). Due to conservation considerations, in the pXRF work only surface analysis was permitted, without the removal of any corrosion, although flat surfaces revealing apparently clean metal were selected for analysis whenever possible. Comparison of the pXRF and OES data for the major elements on the same objects showed that copper was systematically underestimated in the pXRF data, whereas both the tin and the lead were overestimated. Arsenic and silver showed severe surface enrichment as measured by pXRF.

Although it is critically dependant on the instrument used, the calibration methodology and the degree to which the surface can be cleaned, this study showed severe problems with pXRF data on uncleaned metal surfaces, both for the alloying elements (in this case copper, lead and tin), and especially for arsenic and silver. Initially we had hoped that using our more robust interpretational methodology for producing copper groups and alloy types (as described in the following chapters), we might still be able to use such data, and incorporate it within the FLAME database. However, comparison of the trace element and alloy classifications produced using the OES and pXRF on the same set of objects showed that this was not the case. In terms of the Copper Groups (Chapter 4), 44 of the 62 samples are classified differently depending on the analytical method used, whereas for alloy types (Chapter 5), 9 of the 62 are classified differently. This clearly shows that pXRF trace element data collected using the conditions described above are not useable for determining Copper Groups on objects, but major elements may be useable *in extremis* for allocation to alloy types, although subject to considerable uncertainty.

When carefully used, with appropriate standards and calibration, on materials which have, or can be prepared to give, clean representative surfaces, then there is no reason why such instruments cannot give useable results, comparable with other forms of analysis. The onus is on the analyst or the user of the data to demonstrate that the data are reliable, preferably by including the analyses of standard reference materials within the data (although even this does not guarantee the quality of the data if the samples are not adequately prepared). We have included pXRF data in our database when we have encountered it, but it is clearly marked as such. We would not normally include it in our interpretative procedures unless there was conclusive evidence that it was obtained from clean flat metal surfaces, or that suitable sampling protocols had been used, and the appropriate calibration procedures had been applied.

Specific Data Quality Issues relevant to Archaeological Copper Alloys

There are some specific data quality issues in the analysis of archaeological copper alloy objects that need discussing before we describe the interpretative methodologies in detail in the subsequent chapters.

Difficulties of measuring high levels of lead in archaeological copper alloy objects

High levels of lead (>10%) are very difficult to measure accurately in copper alloy objects, for a number of reasons. One is the inherent inhomogeneity of copper alloys containing lead, arising from the insolubility of lead in copper when present at concentrations of more than 1%. The degree of segregation depends on the cooling rate (Staniaszek and Northover 1983), but the result is phase separation into lead-rich and copper rich phases as the alloy cools. The solubilities of lead in the copper phase and copper in lead phase are poorly known, but likely to be very low (Teppo *et al.* 1991). This separation causes the lead-rich phase to concentrate in the last parts of the casting to freeze, since it has the lower melting point. Hughes *et al.* (1982) showed this rather dramatically by sectioning a Bronze Age sword from Selbourne, UK. Using an electron microprobe, they obtained values for Pb as low as 0.26% at the cutting edge, but as high as 12–17% in the centre of the blade, with occasional values up to 50% when the electron beam hit a lead-rich globule. Such extreme examples of segregation might be rare, but could be one explanation for the significant discrepancies between the estimation of lead using different instrumental methods, as noted above, if they produce information from different depths within the object. This also implies that any surface method of analysis, such as X-ray fluorescence, may severely underestimate the lead content of a heavily leaded object. Another complication in the analysis of lead in copper is the fact that the lead-rich phases are much softer then the copper, and are easily pulled out or smeared when polishing a cut section, which also causes the lead to be underestimated in the sample when analysed by electron microscopy or XRF.

Potentially, the phenomenon of segregation calls into question the meaning of any measurement of lead in a leaded object, unless the entire object is dissolved and the lead weighed. It is worth noting that gravimetric analysis nearly did just this, historically using a sample which was probably large enough (~1 g or more) to provide a representative sample. This is why eminent archaeometallurgists of the 1960s such as Earle Caley (1964a) much preferred gravimetric analysis for the major elements to the spectroscopic methods of the time.

Measuring arsenic in the presence of lead by ED-XRF or SEM

In techniques based on energy-dispersive X-ray fluorescence, primarily XRF and SEM, there is a considerable difficulty in measuring arsenic in the presence of lead because of spectral overlap (**Figure 4**). Lead has three major L emission lines—the L_α at 10.5 keV, the L_β at 12.6 keV and the $L\gamma$ at 14.8 keV, with theoretical relative intensities of 100, 70 and 10 respectively. The arsenic lines most frequently measured are the K_α at 10.5 keV and the K_β at 11.7 keV, with theoretical relative intensities of approximately 100:10. Clearly, because of the resolution of most energy-dispersive detectors, the Pb L_α and the As K_α lines are irresolvable, and therefore the emission line measured at 10.5 keV will record both lead and arsenic, if present. This is a severe problem in leaded bronzes, which often have several percent or more of lead, but only a trace of As (<0.1%). There are several possible solutions, but all will result in poorer precision and accuracy for arsenic than for any of the other elements. One possibility is peak deconvolution by Gaussian peak fitting using specialised software (unlikely to be particularly reliable). An even less robust procedure is to estimate the contribution of lead to the combined (Pb + As) peak at 10.5 keV by measuring the Pb L_β peak (12.6 keV) and calculating the Pb L_α as 100/70 of the Pb L_β peak. This can then be subtracted from the combined (Pb + As) peak to give the arsenic intensity. It is to be expected that the error estimate associated with the arsenic concentration derived in this way is likely to be extremely large, and in most cases will exceed the magnitude of the estimated arsenic concentration itself if arsenic is present at trace levels. A better alternative is to simply quantify lead from the Pb L_β peak, and then to use the As K_β to quantify arsenic if lead is present, or to use the As K_α if lead is demonstrably not present. Quantifying lead from the L_β peak is no worse than doing so from the L_α since the two have intensities of the same order of magnitude (providing no other interferences are present). If there is no lead present (or only an insignificant quantity compared to the arsenic), then the arsenic can be quantified from the combined (Pb + As) peak at 10.5 keV, which gives the most precise estimate of arsenic. If lead is present, then it is probably better to quantify it on the interference-free As K_β line, although there is a significant reduction in precision and sensitivity. In a large analytical study of Medieval brass scientific instruments

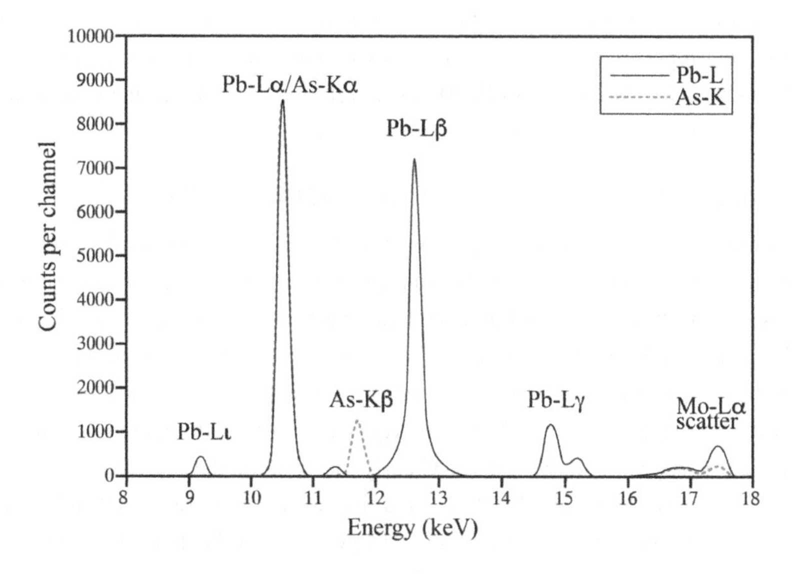

Figure 4:
Overlap between lead and arsenic emission lines in the X-ray spectrum from either XRF or SEM.

using XRF (Pollard *et al.* 2017, 264), the minimum detectable level of arsenic in the presence of lead, measured using the last method described, was calculated to be 0.18%, which has significant implications for the copper group methodology described in the following chapter. Approaches for dealing with the relative lack of sensitivity for arsenic in the presence of lead are discussed in Chapter 4.

How good is 'general agreement'?

The majority of the inter-laboratory comparisons and round-robins carried out to date have concluded that most analytical techniques produce data which show 'general agreement' between methods, with the most common exceptions being either data produced by portable XRF on relatively unprepared metal surfaces, or analysis of lead in heavily leaded bronzes. However, we need to be realistic about what 'general agreement' means. Pernicka *et al.* (1997, fig. 15) compared data on the same objects carried out by OES in Moscow and NAA in Heidelberg, and superimposed upon the figures a band corresponding to 'agreement within a factor of three', implying that this reflected 'general agreement'. This is only notable because it is one of the few cases where authors have attempted to quantify what they feel is acceptable from such a comparison. Mostly, a correlation line is drawn, and as long as it is a positive correlation, it is classified as showing 'general agreement'. What agreement should we expect? It is well known that some if not all archaeological copper alloy objects are chemically inhomogeneous, and that actually this inhomogeneity might limit the analytical reproducibility, even in the most precise of measurements. In some extreme cases, such as the measurement of lead in highly segregated castings (e.g., Hughes *et al.* 1982), we should be grateful if we can replicate the lead to better than an order of magnitude! There is obviously a balance to be struck between highly precise measurements made on extremely small areas (e.g., laser ablation ICP-MS) on the one hand, and much less precise measurements on larger homogenized samples, as was the case in OES or gravimetric analysis. We might have to contemplate the idea that these older analyses are actually more representative of the bulk composition than some of the highly precise modern analyses. Realistically, it is likely that Pernicka *et al.*'s implied definition of 'general agreement' being within a factor of three is a good working criterion at least for the trace elements in copper alloys. One would like to think that for the major elements we should be able to do better, and in general we do, but the example of segregated lead counsels some caution in this matter. However, even if we adopt these assumptions about 'general agreement', the statement by Cuénod (2013, 146) quoted above is still critical. When using parametric numerical methods to cluster or differentiate between clusters (or methods which standardize data by using the mean and standard deviation to create 'standard scores'), we run the very strong risk that the dissimilarities will be dominated by systematic differences between analysts or analytical methods. In other words, the resulting classifications may reveal more about the analytical methods used rather than any inherent differences between samples. It is our assertion, as explained in the following chapters, that when dealing with complex data sets which might only show 'general agreement', it is better to avoid such methods.

'Analytical hygiene'

The term 'analytical hygiene' is used here in the same sense that 'chronometric hygiene' (Spriggs 1989) has been applied to issues of quality assurance (QA) in radiocarbon dating for archaeology, where it is intended to specify best practice in the reporting of data, and to provide some guidance as to how to interpret published data. What is ideally required from the publications containing analytical data is the routine availability of quality assurance data (QA) from each laboratory involved in the chemical and isotopic analysis, either published with the data themselves, or available online. Many labs will keep such information as logs, but until recently most academic journals have had little capacity to publish such data, although online publishing of supplemental data now makes this feasible. In archaeology, the radiocarbon laboratory community have set the standard for this sort of protocol, with the publication of a series of regular inter-laboratory comparisons, and QA data for individual laboratories being available online or on request. Ideally, publications of analytical data of any sort should include sufficient data for the reader to gain an understanding of the quality of the data, in terms of recognized parameters such as *accuracy*, *precision*, *detection limits*, etc., as relevant to the type of analysis. In analytical chemistry this requires publication of a considerable amount of information on the standards used to calibrate and validate the samples, as well as the order in which samples were run, and the degree of replication, and so on. It is now increasingly common for journals to insist that all the raw data are published, and often quality assurance data can be included in online appendices.

Appendix: Idealised "analytical hygiene" for the analysis of archaeological non-ferrous metals

The appendix to this chapter suggests a highly idealized list of the information which would be of value when publishing analytical data on archaeological bronze objects. A full discussion of quality assurance (QA) issues is given in Pollard *et al.* (2007, 306–321), but a brief definition of the relevant terms is summarized here.

Precision is the degree of reproducibility between replicate measurements on the same sample. *Within-run* precision means the degree of reproducibility obtained on replicates during the same run of the instrument. Variation of precision between runs is usually greater, and is sometimes termed the *repeatability* of a measurement.

Accuracy refers to the closeness of the experimental result to the 'true' value. Precision is, therefore, an internal measure of analytical quality, whereas accuracy relates to an absolute measure.

Precision can be evaluated from the repeated analysis of any material which is included in the analysis as an unknown, and which is sufficiently close in composition to the samples being analysed. Often, an *in-house standard* can be used to determine precision, since it is not necessary to know the composition of this standard, other than to know that it is reasonably close in composition to the unknown material. These can be cheap, homogeneous and readily available materials which can be repeatedly analysed without incurring excessive costs, as would be the case if international reference materials were used. *Accuracy*, on the other hand, can only be determined relative to the values of an internationally recognised standard reference material.

Most instrumental methods of analysis require the use of *standards*, either to calibrate the instrument, or to provide quality assurance data, or both. It is important to appreciate that the term 'standard' is used to describe a range of materials fulfilling very different purposes, as follows:

Calibration standards

Almost all methods of chemical analysis require a series of calibration standards containing different amounts of the analyte in order to convert instrumental readings into absolute concentrations. These can be synthetic solutions (for liquid-based analyses), pure elements or compounds (for SEM, microprobe and

XRF), or primary multi-element standards with known concentrations of the relevant elements. They should be used purely for calibrating the instrument, and not used for QA purposes, unless they are specifically analysed as 'unknowns' independently of the calibration process.

Primary standards

These are internationally agreed Standard Reference Materials (SRMs), also called Certified Reference Materials (CRMs), which have been accurately and repeatedly analysed, often using different techniques, in different laboratories. The results are fully documented and certified, although not always for all required elements. Such standards are commercially available, but are often expensive (and, therefore, too costly to use repeatedly in each analytical run). If primary standards are included in the analytical runs as unknowns, they can be used to assess the *accuracy* of the analysis. In archaeology, it is often useful to use a particular sample or set of samples as primary standards, even if they are not certified, because it may be the only way of getting data on certain elements. Often such samples are only quantified by consensus between participating laboratories, and it should therefore be made clear that they are not certified.

Secondary standards

Because of the cost of certified primary standards, it is normal to use them sparingly, and it is common to use materials which can be included in each analysis which may not be fully certified, but whose values can be related to a primary standard. These are often referred to as *in-house* standards, although they may sometimes be shared between several laboratories. They can be analysed sufficiently often within the analytical run to enable the calculation of the precision of the analysis, but are less useful for accuracy.

Minimum detectable level (MDL)

Quoting the minimum detectable level, or detection limit, for each element analysed is one of the most important issues in the QA of analytical data. It can be defined in several ways, but the simplest is in terms of being the concentration of a particular element in a sample which produces an analytical signal equal to twice the square root of the background above the background. This definition corresponds to the 95% confidence interval, which is adequate for most purposes. Higher levels, such as 99%, can be defined by using a multiplier of three rather than two if required. The MDL will vary from element to element, from machine to machine, and even from day to day, and should be calculated explicitly for every element each time an analysis is performed. It can be calculated most straightforwardly from the data generated by the linear regression used to calibrate the machine for each element (see Pollard *et al.* 2007, 319).

Idealized QA data

The following is intended to describe the IDEAL set of data to be produced for the analysis of ancient metal. It is generalized to allow for a wide range of types of analysis, so not every point is relevant to all procedures.

Stages

1. Sample Description
   a. Full details of archaeological object (typology, size, weight, ...), including photograph and/or drawing
   b. Full details of archaeological context (location, type of context, ...), including photograph and/or drawing
   c. Chronological details, including how dated (from typology, context, or direct dating by radiocarbon)
   d. Current location (accession number, details of any publications, drawings or photographs)
2. Sampling Details (if a sample is taken)
   a. Where sampled (on object), including drawing or photograph to show sample location
   b. How sampled
   c. Description of sample
   d. Location of sample (if any left) after analysis
3. Sample Preparation
   a. Solid sample – has it been pelletized, or mounted, prepared and polished?
   b. Has a metallographic record been made?
   c. Liquid sample – dissolution procedure, concentrations used
4. Analytical Method
   a. Description of instrument
   b. Location of instrument
   c. Operating conditions
   d. Number of measurements taken per analysis
   e. Is the analysis 'bulk' or 'surface'?
   f. In heterogeneous samples, how many phases have been analysed?
   g. Levels of detection, precision, accuracy
   h. Operator, date of analysis, description of running order of samples logged

5. Analytical Protocols
   a. Primary standards
   b. Secondary standards
   c. Calibration Procedure
   d. Normalization
6. Description of Data
   a. Elements measured
   b. Isotopes measured
   c. Analytical total
7. Publication of data to include all of the above!

Chapter 4

Trace Elements and 'Copper Groups'

This chapter describes how we manipulate the trace element data from an assemblage of copper alloy objects to produce 'Copper Groups', and how we then carry out mapping and profile analysis to interpret these data. We also discuss how in some cases the Copper Groups can be used to detect the recycling of metal, based on the thermodynamic properties of the different trace elements. As pointed out in Chapter 2, we focus on four trace elements, namely arsenic (As), antimony (Sb), nickel (Ni) and silver (Ag). The reasons for this choice are primarily practical. Firstly, most chemical analyses of copper alloys that include trace element data will report these four trace elements. This allows us to include the largest possible number of samples when calculating Copper Groups. Secondly, as demonstrated below, these four elements cover a range of thermodynamic behaviours in molten copper—under oxidising conditions, arsenic is volatile, antimony slightly less volatile, and silver and nickel stable. The choice of only four elements is also distinctly practical. All possible combinations of presence/absence for four elements (defined in **Table 1**) give 16 possible 'Copper Groups'. If we were to use five elements, then there would be 32 possible Copper Groups, which would be cumbersome to display and interpret. We do not claim that the four elements used here are the only ones to carry useful information—plainly not. Where reported, for example, bismuth (Bi) or cobalt (Co) can be extremely useful in distinguishing between copper from different sources. However, the combination of the overall availability of data for the four selected elements, and the potential complexity of using five or more elements, has resulted in our standard starting point being the use of arsenic, antimony, silver and nickel. As emphasised in Chapter 2, our primary aim is to characterise *change* in the material record rather than specifically identify provenance, and this is usually possible using this combination of four trace elements without considering further elements. We can imagine that, for certain data sets, there might be an advantage to swapping, say, nickel for bismuth, providing the data will allow it, but this immediately loses the *universality* of our approach, which we see as a distinct advantage when carrying out large-scale comparisons, as discussed below. We do, however, sometimes consider splitting Copper Groups defined on the basis of the four trace elements according to the presence or absence of a fifth element. We might think, for example, of a copper group defined as NYYY (CG13: Cu +Sb, Ag and Ni: see below) being split into further sub-groups defined by the presence or absence of a fifth element. This would again, however, lose the benefit of universality, and it is fair to point out that in the many case studies carried out to date we have not yet felt the need to use a fifth element to aid in this preliminary classification step.

CG	As	Sb	Ag	Ni	Code (AsSbAgNi)	Description	Shorthand
1	<0.1%	<0.1%	<0.1%	<0.1%	NNNN	Clean copper	Clean
2	>0.1%	<0.1%	<0.1%	<0.1%	YNNN	Cu + As	CuAs
3	<0.1%	>0.1%	<0.1%	<0.1%	NYNN	Cu + Sb	CuSb
4	<0.1%	<0.1%	>0.1%	<0.1%	NNYN	Cu + Ag	CuAg
5	<0.1%	<0.1%	<0.1%	>0.1%	NNNY	Cu + Ni	CuNi
6	>0.1%	>0.1%	<0.1%	<0.1%	YYNN	Cu + As, Sb	CuAsSb
7	<0.1%	>0.1%	>0.1%	<0.1%	NYYN	Cu + Sb, Ag	CuSbAg
8	<0.1%	<0.1%	>0.1%	>0.1%	NNYY	Cu + Ag, Ni	CuAgNi
9	>0.1%	<0.1%	>0.1%	<0.1%	YNYN	Cu + As, Ag	CuAsAg
10	<0.1%	>0.1%	<0.1%	>0.1%	NYNY	Cu + Sb, Ni	CuSbNi
11	>0.1%	<0.1%	<0.1%	>0.1%	YNNY	Cu + As, Ni	CuAsNi
12	>0.1%	>0.1%	>0.1%	<0.1%	YYYN	Cu + As, Sb, Ag	CuAsSbAg
13	<0.1%	>0.1%	>0.1%	>0.1%	NYYY	Cu + Sb, Ag, Ni	CuSbAgNi
14	>0.1%	>0.1%	<0.1%	>0.1%	YYNY	Cu + As, Sb, Ni	CuAsSbNi
15	>0.1%	<0.1%	>0.1%	>0.1%	YNYY	Cu + As, Ag, Ni	CuAsAgNi
16	>0.1%	>0.1%	>0.1%	>0.1%	YYYY	Cu + As, Sb, Ag, Ni	CuAsSbAgNi

Table 1:
The definition of Copper Groups.

We see a significant advantage in using, at least initially, the same set of elements (and the same cut-off values) for all analyses—that of *universality*. This means that the calculated Copper Groups from, for example, the Caucasus, can be directly compared with those from Mongolia. In a project such as FLAME, which considers Bronze Age Eurasia to be a set of interlinked metal systems, it is extremely important to be able to compare across space and time. We suggest that, as described in Chapter 1, one of the major limitations of previous large-scale studies of copper alloy chemistry is that they have all devised their own classification systems for separating objects on the basis of trace element composition (e.g., the decision trees of SAM, and the classifications proposed by Northover). Such classifications might well be effective for the purposes of the particular project in hand, but they are likely to only be applicable to the data included within each dataset, and will therefore differ from project to project, depending on who has done them. Thus there is no way that the resulting classifications can be directly compared from one study to another. Nor can new data be easily added to existing datasets, since the classificatory cut-offs are often defined from an analysis of the distribution of particular elements within the dataset, either using some assumed normality of overlapping distributions, or looking for 'natural breaks' in the data. Thus there are clear advantages to universality, even when considering data on a scale smaller than the entire Eurasian continent. It must be emphasised, however, that in claiming universality we are simply referring to the application of a common preliminary sorting methodology, and we do not imply that a common interpretation can be applied to the derived datasets irrespective of social context. We completely appreciate that an interpretation of metal movement or relationships in a simple small-scale social context may be completely inapplicable to a large-scale centralised economy. Recycling, for example, may be a dominant factor in one context but be completely irrelevant in another.

Deriving Copper Groups

In order to convert the chemical data into Copper Groups, we first need to define the membership of one or more assemblages on the basis of chronology, typology, time or space, depending on the nature of the question being investigated. It is important to check that the data for the selected objects are as 'clean' and complete as possible, as described in Chapter 3 —errors corrected, non-numeric symbols removed or dealt with in a systematic and transparent way, and, as far as possible, the validity of the data checked. The next step is to carry out a simple presence/absence classification system based on the four trace elements discussed above—arsenic, antimony, silver and nickel—for all of the objects in each of the defined assemblages. This is a simple heuristic sorting step, which allows us to see the trace element characteristics of each assemblage, and to compare the dominant chemical signals running through the different assemblages (Bray 2009; Bray and Pollard 2012; Bray *et al.* 2015). The presence or absence of these four trace elements is *most likely* to be related to ore-source, since they tend to be either present or absent in the ores known to have been used in antiquity, but we make no assumptions at this stage about allocating a particular copper group to a specific ore source, known or unknown. A single ore source could produce copper classified into more than one copper group, and, conversely, copper of a single group could come from more than one mine. At this stage, we are only interested in the geographical, typological or chronological patterns that are revealed by this process, with no prior assumptions required about mines or geology. Tracing these changes over time, through a landscape, or between social contexts, is at the heart of interpreting metal flow.

Before allocating an object to a copper group, for each sample we mathematically remove the alloying elements and renormalize the analysis to 100%, since we assume at this stage that the trace elements are associated primarily with the copper. If this assumption is true, then the addition of substantial quantities of alloying elements will simply dilute the trace element values, and may lead to misclassification. We take the alloying elements in most copper objects to be lead, tin and zinc, although of course not all three are always present, and, further, at this stage we assume that the alloying elements do not carry with them any of the selected trace elements. This is most likely not to be a safe assumption in the case of silver and lead, which is discussed in more detail below. The possible use of arsenic (or less commonly antimony) as a deliberately added alloying element rather than a trace 'contaminant' is discussed below. The formula used for the recalculation of the trace element concentrations without the alloying elements is as follows:

$$[x]_{corr} = [x]_{meas} * 100/(Tot - [Sn] - [Pb] - [Zn])$$

where $[x]_{corr}$ is the corrected concentration of the trace element, $[x]_{meas}$ is the measured concentration of that element, *Tot* is the original analytical total, and [Sn], [Pb] and [Zn] are the measured concentrations of tin, lead and zinc. There may be an issue if the original analytical total (*Tot*) is significantly different from 100%, which could indicate either a poor analysis, or that one or more elements have been missed or poorly determined. It is probably safest to omit completely such an analysis from the definition of the assemblage. A useful rule of thumb is to omit any analyses with an analytical total <95% or >105%). If this is not done, then it is probably best to replace the 100 in the above formula with the actual analytical total (*Tot*), in order to avoid adding more uncertainty into the values. In certain sets of analyses (for example, the SAM data), copper was often not directly measured in the original publication. Instead, it was estimated by the difference between the total of the measured elements from 100%. This does not numerically affect the above procedure, but in such cases the reported analytical total is artificially set to 100%, and cannot therefore be used to indicate the quality of the overall analysis. The same is true for analytical procedures which automatically normalize the data to 100% as part of the calibration process. The analytical total is the single best indicator of the reliability of any analysis, so those results which have a genuine estimate of the true total are inherently preferable to those which do not (providing that the estimate is close to 100%!).

After removing the alloying elements and renormalizing, we then allocate the composition of each object in the assemblage to one of 16 categories, on the basis of presence/absence (Y/N) of each corrected trace element value ($[x]_{corr}$), using the definitions listed in Table 1. The categories are arbitrarily labelled and ordered according to a set of rules, based on the presence/absence of the elements in the fixed order of As/Sb/Ag/Ni. Thus, an object with arsenic above 0.1% but everything else below 0.1% would be labelled 'YNNN' and assigned to Copper Group 2 (CG2). As noted above, for most datasets, we use a figure of 0.1% for $[x]_{corr}$ as the division between presence and absence. This figure is essentially a compromise which allows us to use as much data as possible from differing sources, thereby allowing us to create assemblages of the maximum possible size. As discussed in Chapter 3, different analytical techniques can have radically different detection limits for a particular element, as well as each analytical technique having different limits for each element, and sometimes different limits for the same element within different major element matrices. For example, one might expect 0.01% (or even 0.001% and below) to be the minimum detectable level (MDL) for more modern analytical techniques such as inductively coupled plasma mass spectrometry (ICP-MS) when measuring trace elements in copper, whereas in older methods, such as optical emission spectroscopy, it may be 0.1% at best. When we are dealing with large datasets containing either data from mixed analytical sources, or only data from older methods, we have to use a cut-off relevant to the poorest quality data in the dataset, or else reject those data. Clearly, using a cut-off that is much higher than the minimum detectable level of the best data is not damaging to the classification outcome (although there is obviously the risk of missing some details present in the data), whereas using one much lower than that of the poorest data runs the risk of interpreting what is essentially noise. Arguably, if the assemblages contain only high quality (i.e., modern) data, then a case could be made for reducing the cut-off values for that particular exercise. Normally we would not do this, in the interests of retaining universality, as discussed above, but for some specific analyses it might be a useful additional procedure. There is a particular problem when using data generated by energy dispersive X-ray methods (XRF and SEM) on arsenic in the presence of lead (see previous chapter), since the MDL for arsenic in such cases might be as high as 0.18%. In these circumstances any values recorded as being below 0.18% are unreliable, and therefore the presence/absence allocation with a 0.1% cut-off is also unreliable for arsenic. Ways of dealing with this situation are discussed below.

After allocating each object in an assemblage to one of the sixteen categories described above, the next step is to summarise the combined results. The number of objects in each copper group is expressed as a percentage of the total number of objects in the assemblage. We refer to this percentage as the *ubiquity* of that copper group. Thus, if all the objects in an assemblage are categorized as Copper Group 2, then this would result in an assemblage with CG2 = 100%, and all other Copper Groups equal to zero. A random distribution of objects between all possible Copper Groups would have each Copper Group reported as 6.25%. The pattern for each assemblage is most simply summarised numerically as a single row of percentages in a table of Copper Groups. When comparing a set of assemblages, they can be entered as successive rows in the table, and patterns are clearly seen simply by comparing the changes between rows (see for example, **Table 2**). A useful practice is to colour code the percentages, with a scheme such as red signifying ubiquities above 50%, amber between 30% and 50%, and yellow for 20-30%, depending on the range within the data. In this way significant components can be immediately identified, and changes down the table can be easily evaluated. The same data can also be displayed graphically as a pie chart representing the composition of a single assemblage (**Figure 1**). Several pie charts can then be displayed on a map, allowing variations across space to be visualised.

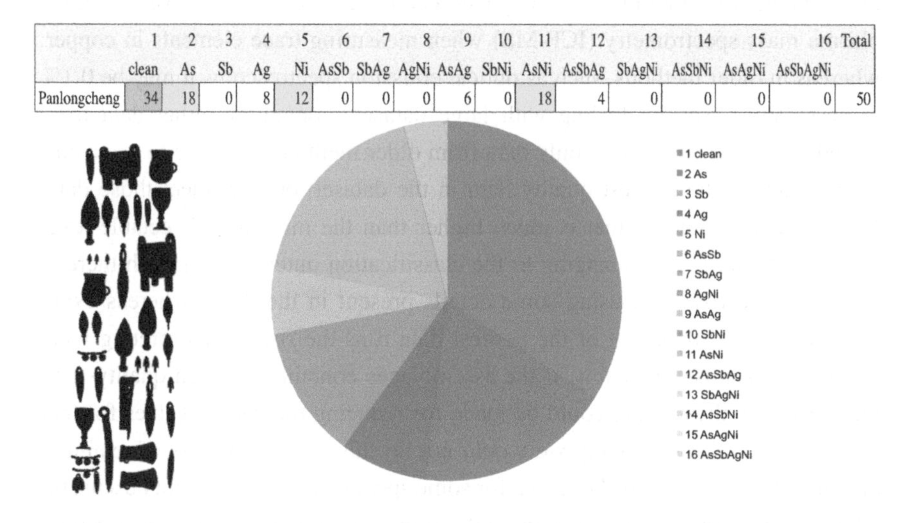

	1 clean	2 As	3 Sb	4 Ag	5 Ni	6 AsSb	7 SbAg	8 AgNi	9 AsAg	10 SbNi	11 AsNi	12 AsSbAg	13 SbAgNi	14 AsSbNi	15 AsAgNi	16 AsSbAgNi	Total
Panlongcheng	34	18	0	8	12	0	0	0	6	0	18	4	0	0	0	0	50

Figure 1.
A hypothetical example showing the ubiquity of Copper Groups in an assemblage of 50 objects, presented as a table and a pie chart.

An example of such a table is shown in Table 2, where the ubiquity of Copper Groups for Chinese metal objects from the Erligang (Zhengzhou) period (c. 1500–1400 BCE), the Shang (Anyang) Dynasty (c. 1400–1046 BCE) and the Western Zhou Dynasty (1046–774 BCE) are shown sequentially. At the very broadest level, this shows that the Erligang period metalwork contains 18% of objects assigned to

	1 clean	2 As	3 Sb	4 Ag	5 Ni	6 AsSb	7 SbAg	8 AgNi	9 AsAg	10 SbNi	11 AsNi	12 AsSbAg	13 SbAgNi	14 AsSbNi	15 AsAgNi	16 AsSbAgNi	Total
Erligang	26.3	13.2	0	18.4	2.6	0	0	0	23.7	0	2.6	10.5	0	2.6	0	0	38
Shang	19.8	26.7	1.4	3	0	5.1	1.1	0.2	17.2	0.2	0.7	21.6	0.2	0.2	1.1	1.4	435
Western Zhou	15.5	30.5	0.9	4.1	0	6.8	0	0.5	18.2	0	0	21.8	0.5	0	0	1.4	220
Panlongcheng	34	18	0	8	12	0	0	0	6	0	18	4	0	0	0	0	50

Table 2:
An example of the use of a table of ubiquities of Copper Groups (see text).

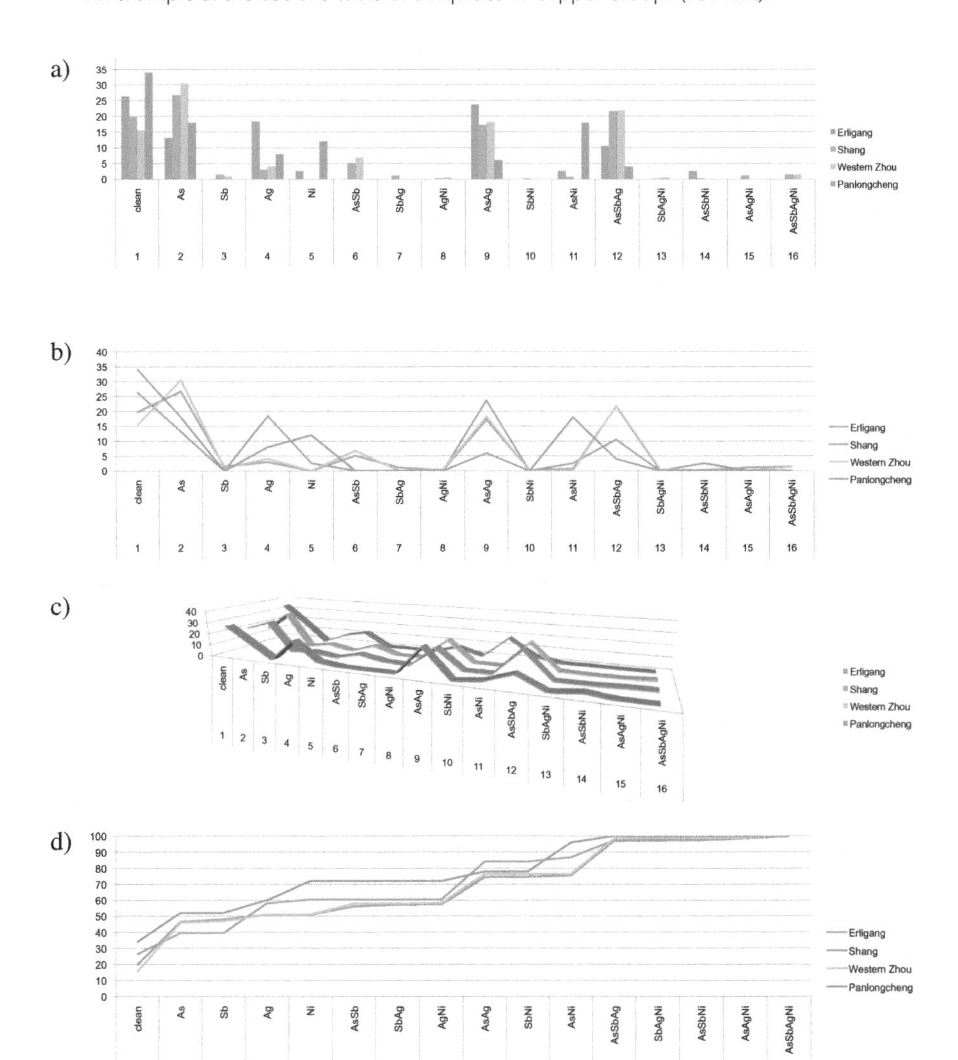

Figure 2:
Alternative graphical representations of the data given in Table 2. a) bar chart, b) skyline plot, c) ribbon plot, d) cumulative frequency curves.

CG4 (Cu + Ag), which is absent from the later periods. It also shows the relative similarity between the CGs used during the Shang and Western Zhou dynasties, which we have interpreted as continuity in the sources of copper used between the Shang and Western Zhou, but that some changes occurred between Erligang and Shang (Pollard *et al.* 2017a). The same table also shows the ubiquity of Copper Groups at the site of Panloncheng, which is an Erligang period settlement on the Yangzi River. The difference between Erligang (Zhengzhou), which represents the metal in circulation at the Erligang capital, and Panlongcheng (e.g., the presence of CGs 5 (Cu + Ni) and 11 (Cu + As, Ni) at Panlongcheng) suggests that there is not a simple correspondence between the metal in circulation at these two sites (Liu *et al.* 2017; in prep.).

Figure 2 shows a series of graphical representations of the data shown in Table 2. The bar chart (Fig. 2a) allows a direct comparison of the ubiquity of each assemblage within the 16 Copper Groups. Figure 2b is also a histogram, but the top of each column has been joined together for each assemblage to give a 'skyline plot'. The 'ribbon plot' (Fig. 2c) is the same presentation rendered in three dimensions. The fourth plot (Fig. 2d) is a cumulative frequency curve of the data presented in Table 2, which perhaps shows most directly the similarity between Shang and Western Zhou, and the different nature of Erligang and Panlongcheng. Cumulative frequency curves are discussed in more detail in the next chapter.

All these graphical representations are equivalent, and are based on the arbitrary sequence of Copper Groups defined in Table 1. Such comparisons are valid between assemblages provided this sequence is adhered to. Obviously, any presentation of the type shown in Figure 2 will become difficult to interpret if the number of assemblages in the comparison becomes too great—generally, four is about the maximum.

Exploring the effect of the choice of trace element cut-off value

Given that the selection of a cut-off value of 0.1% between presence and absence for the trace elements is a compromise designed to allow the use of the maximum number of analyses, it is important to consider whether the use of this figure is affecting the interpretation of the data. As noted in Chapter 3, within the chemical data the minimum detectable level will vary from element to element and from instrument to instrument. For some elements by certain analytical methods, the cut-off might be above this value, and the allocation of individual samples to a specific Copper Group could be unstable. One way of checking the robustness of the allocation to Copper Groups is to systematically vary the cut-offs, compile a summary table for each cut-off value as described above, and compare the stability of the ubiquities of each group as the cut-off varies. The FLAME database software, as described in Chapter 7, allows the cut-off to be set by the user independently for all four trace elements. **Table 3** shows an example from a dataset consisting of 298 objects from the Sackler Collection (Bagley 1987), measured by AAS for major elements and NAA for traces. The table shows how the allocation to Copper Group varies as the cut-off for all four trace elements is increased step-wise from 0.01% to 0.2%. **Figure 3** shows the changes in the allocations for CGs 1 (Cu only), 9 (Cu + As, Ag), 12 (Cu + As, Sb, Ag), 16 (Cu + As, Sb, Ag, Ni), which are the major CGs shown in Table 3.

Cut-off (%)	1	2	3	4	5	6	7	8	9	10	11	12	13	14	15	16
	clean	As	Sb	Ag	Ni	AsSb	SbAg	AgNi	AsAg	SbNi	AsNi	AsSbAg	SbAgNi	AsSbNi	AsAgNi	AsSbAgNi
0.01	1.0	0.3	0.0	0.0	0.0	0.0	0.3	0.3	13.1	0.0	0.0	47.8	0.0	0.3	4.0	32.7
0.02	1.7	2.4	0.0	1.7	0.3	0.3	0.7	0.0	22.9	0.0	0.0	47.1	0.0	0.3	3.0	19.5
0.03	3.7	3.7	0.0	3.0	0.0	0.7	0.3	0.0	29.3	0.0	0.7	43.8	0.0	0.3	2.0	12.5
0.04	5.1	6.4	0.0	3.7	0.0	0.7	0.3	0.0	31.6	0.0	0.7	41.8	0.0	0.3	1.3	8.1
0.05	7.7	8.8	0.3	5.1	0.0	1.3	0.0	0.0	30.3	0.0	1.0	38.7	0.0	0.0	1.7	5.1
0.06	9.1	11.8	0.3	6.1	0.0	1.7	0.0	0.0	29.0	0.0	1.0	35.7	0.0	0.7	1.3	3.4
0.07	11.8	14.8	0.3	5.7	0.0	4.0	0.0	0.0	26.3	0.0	1.3	32.0	0.0	0.7	0.0	3.0
0.08	16.2	20.5	0.3	3.7	0.0	5.1	0.0	0.0	21.2	0.0	1.0	29.3	0.0	0.7	0.0	2.0
0.09	19.2	22.2	0.7	4.0	0.0	6.7	0.0	0.0	18.9	0.0	0.7	24.9	0.0	0.7	0.0	2.0
0.10	24.2	24.2	0.7	1.7	0.0	6.4	0.0	0.0	17.8	0.0	0.3	22.2	0.0	0.3	0.0	2.0
0.11	25.3	29.3	0.7	1.7	0.0	7.4	0.3	0.0	14.1	0.0	0.3	19.5	0.0	0.0	0.0	1.3
0.12	27.6	31.3	0.7	1.3	0.0	7.1	0.3	0.0	13.1	0.0	0.3	16.8	0.0	0.0	0.0	1.3
0.13	29.6	33.0	0.7	1.3	0.0	8.1	0.0	0.0	11.1	0.0	0.3	14.5	0.0	1.3	0.0	0.0
0.14	31.3	35.0	0.7	1.7	0.0	7.7	0.0	0.0	8.8	0.0	0.3	13.1	0.0	1.3	0.0	0.0
0.15	33.7	34.0	0.7	1.3	0.0	9.1	0.3	0.0	8.1	0.0	0.3	11.1	0.0	1.3	0.0	0.0
0.16	35.7	36.0	1.0	0.3	0.0	8.1	0.0	0.0	7.1	0.0	0.3	10.1	0.0	1.3	0.0	0.0
0.17	37.7	35.7	1.0	0.3	0.0	8.1	0.0	0.0	6.1	0.0	0.3	9.4	0.0	1.3	0.0	0.0
0.18	39.4	36.0	1.0	0.3	0.0	8.8	0.0	0.0	5.1	0.0	0.3	7.7	0.0	1.3	0.0	0.0
0.19	40.4	37.0	1.3	0.3	0.0	7.4	0.0	0.0	4.0	0.0	0.3	7.7	0.0	1.3	0.0	0.0
0.20	43.1	35.4	1.7	0.3	0.0	6.7	0.0	0.0	3.4	0.0	0.3	7.7	0.0	1.3	0.0	0.0

Table 3:
Allocation of an assemblage of 298 objects from the Shang bronzes in the Sackler collection (Bagley 1987) to Copper Groups as the cut-off value is varied.

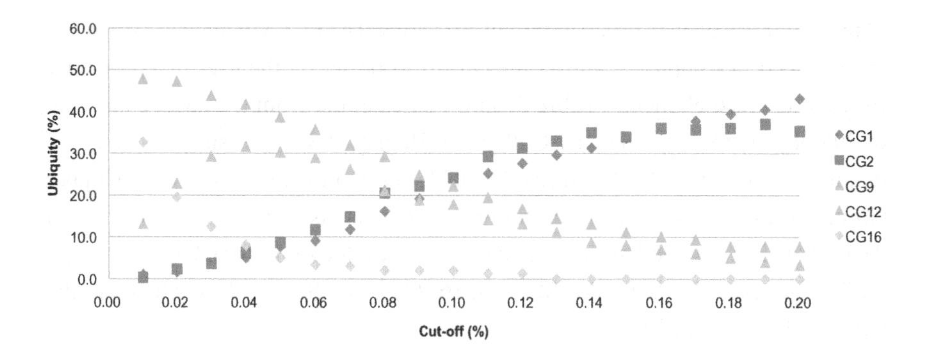

Figure 3:
Variation in allocation of objects to Copper Groups 1, 2, 9, 12 and 16 from Table 3.

At first sight, this does not show any stability in the allocation of the data to Copper Groups. However, it must be remembered that as the cut-off is raised incrementally, there will be a gradual drift away from CGs 2-16 towards CG1 (clean copper) providing the trace elements are within the range of variation of the cut-off, ultimately ending with all objects being classified as CG1. Thus, CG1 should increase at the expense of all other groups as the cut-off rises. With this in mind, Figure 3 clearly shows that with a cut-off set at below 0.05% the allocation to Copper Groups is unstable, but above this figure the attributions behave as expected. What is significant is that CG9 and CG12, whilst both decline above 0.05% because of the increase in allocations to CG1, have approximately the same ratio to each other for any value of the cut-off, and CG16 is shown to be relatively unimportant above 0.05%. This would suggest that, for this type of data, any value of the cut-off above 0.05% will give stable results. That is not to say that the detailed allocation to Copper Groups does not depend on the value of this cut-off, but it does mean that *comparisons between assemblages* will not be affected by the choice of cut-off, provided the same set of cut-offs are used.

A similar pattern is seen when using an assemblage of data produced by electron microprobe (EPMA). Figure 4 shows the variation in the major Copper Groups present in the data from Hanzhong (207 analyses: Chen 2009). This again shows instability in the allocations below 0.05%, but relative stability above that point. As expected, CG1 increases as the cut-off rises, but in this case CG2 (Cu + As) also increases towards the maximum cut-off, at the expense of CG16 (Cu + As, Sb, Ag, Ni). This implies that the samples allocated to CG16 contain low levels of antimony, silver, and nickel, but relatively higher arsenic, so as the cut-off increases samples are gradually re-allocated from CG16 to CG2 and not to

CG1. It is to be expected that were the cut-off to be increased above 0.2%, CG1 would grow at the expense of CG2, up to the maximum value of arsenic in the assemblage.

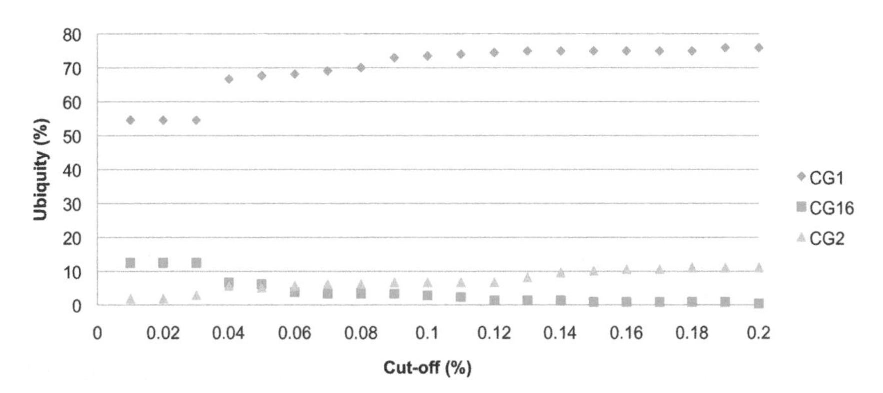

Figure 4:
Variation in the major Copper Groups present as a function of cut-off value in the EPMA data from Hanzhong (207 analyses: Chen 2009)

Both of these examples show that the allocations to Copper Groups behave as predicted above a value of 0.05%, and suggest that 0.1% is a reasonable value to use—although it could be lower. This is to be expected since the trace elements in Figure 3 were measured by neutron activation analysis, and in **Figure 4** by electron microprobe—both of which should have minimum detectable levels for these four elements below 0.05%. As might also be expected, however, data produced by older methods of analysis, especially optical emission spectroscopy (OES), show a pattern that is somewhat more difficult to interpret. **Figure 5** shows the variation in copper group ubiquities for a heterogeneous assemblage of 250 Chinese bronzes from the British Museum, analysed by OES in Oxford in the late 1950s but only published in 2018 (Pollard *et al.* 2018). In this instance the cut-off variation has been extended to 0.45. This figure displays the variation in the major Copper Groups in the data (CG1 (clean copper), CG2 (Cu + As), CG6 (Cu + As, Sb) and CG16 (Cu + As, Sb, Ag, Ni)), and shows a dominance of CG16 up to about 0.15%, but after that CG1 rises steadily. CGs 2 and 6 both rise to a plateau around 0.2%, and then drop away as the cut-off increases. It is not obvious to see where the cut-off should be in this case—it should probably be above 0.15%, but this pre-supposes that CG16 is only registered as a result of random variation in the trace element concentrations, and is not actually present in the data. By looking at some of the less abundant Copper Groups within this dataset, we can investigate this further. **Figure 6** shows the variations in the minor CGs present in the same data as represented in Figure 5. It again shows that with a cut-off below 0.1% the attributions are highly unstable, but that above 0.1% the pattern has settled down. This might suggest that a useable cut-off for older OES data might be 0.1%, but an argument could be made for using a cut-off somewhat closer to 0.15% or even 0.2%. In the interests of universality, we generally apply 0.1% in all cases, but would strongly suggest that anybody carrying out an analysis using this approach should vary the cut-off systematically on their specific dataset before deciding on a value. It is also possible to have different cut-offs for different elements. In specific datasets containing older data, we recommend looking at the trace element distribution profiles within the dataset (see below) before selecting cut-off values. For example, in her DPhil thesis on Iranian metalwork, Cuénod (2013, 273) felt that cut-off values of 0.25% for arsenic, 0.05% for antimony and silver, and 0.1% for nickel were the most appropriate for her particular dataset. There is therefore a balance to be struck between the simplicity of using a uniform set of cut-offs and the quality of the data under consideration.

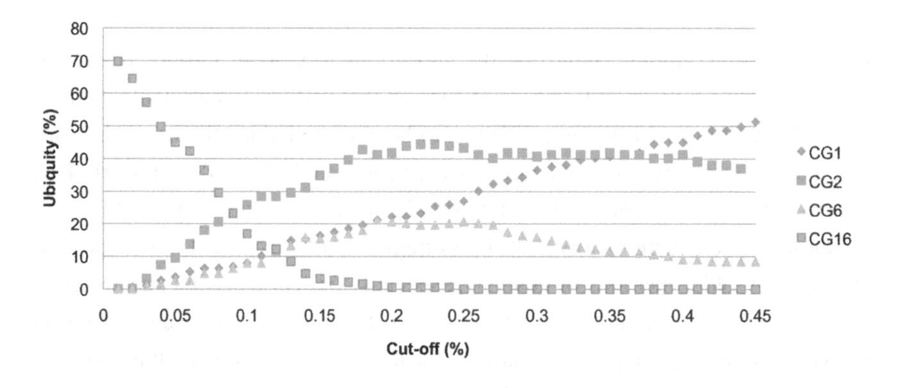

Figure 5:
Variation in the major Copper Groups present as a function of cut-off value in OES data from Chinese and other bronzes in the British Museum (250 analyses: Pollard *et al.* 2018).

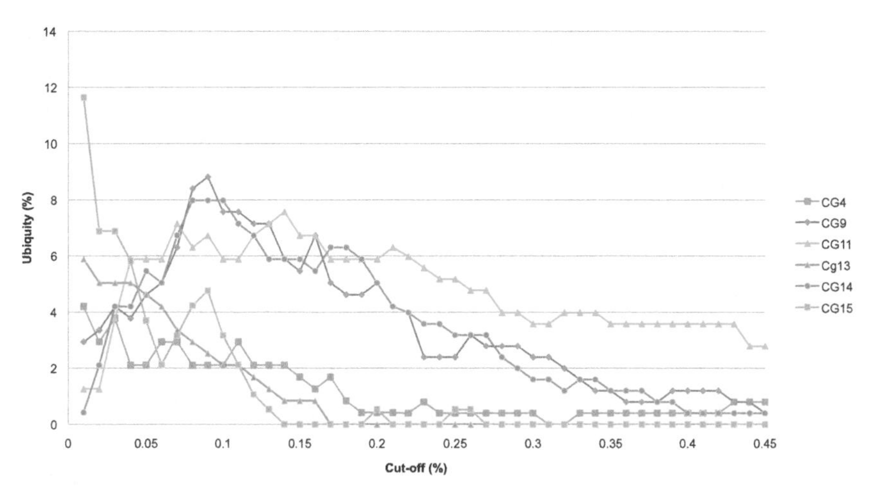

Figure 6:
Variation in the minor Copper Groups present as a function of cut-off value in OES data from Chinese and other bronzes in the British Museum (250 analyses: Pollard ***et al.*** 2018).

Profile analysis

Following this preliminary classification stage, the next step is to look at the distribution profiles for specific chemical elements within the Copper Groups (*profile analysis*). This is discussed in more detail in the next chapter in the context of alloying, but here it is used to gain a better understanding of the characteristics of each copper group, and the relationship between them within the defined assemblage. The profile studied could be that of the (corrected) element used to define the copper group (such as arsenic for CG2), but it could also be the profiles of the other trace elements (such as nickel in CG2, which will all be, by definition, <0.1%). It could also be a profile of the alloying elements for all the objects defined as belonging to a particular Copper Group, or even profiles of the elements not included in the classification process, such as bismuth.

In the case of the profile of trace elements included in the classification (As, Sb, Ag, Ni), particular attention has to be paid to trace element distributions which straddle the cut-off value. For example, if in one assemblage (**Figure 7a**) the arsenic distribution is largely above 0.1%, but has a 'tail' towards lower values, then some objects within the assemblage will be assigned to CG1 (clean copper), but others, which are actually part of the same distribution, will be assigned to CG2 (Cu + As). Another assemblage (**Figure 7b**) might have two discrete distributions, one below the 0.1% cut-off and one above it. Both assemblages would therefore be recorded as containing both CG1 and CG2, but a plot of the arsenic distribution in the two cases will show clearly that the distribution in the first assemblage is continuous, and that the process of allocation to Copper Groups has artificially split this single distribution into two. In the second example, the distribution profile will show two distinct groups. The fact that these two profiles are different but the copper group assignments appear superficially similar does not, however, invalidate the process. In both cases, it indicates that both 'clean' copper (CG1) and Cu + As (CG2) are present in the metal flow, which is the aim of the preliminary classification process. It *would* be a problem, however, if it were assumed that CG1 represented a source of copper that contained no arsenic, and CG2 derived from one that did, but we emphasise again that these Copper Groups do not necessarily correspond to specific 'sources' or 'mine sites'. It simply means that these particular assemblages contain objects classified as both CG1 and CG2, which, in the first example, profile analysis has shown to be part of the same distribution of arsenic. The artificial division between CG1 and CG2 in this case makes no difference to the overall characterization of the assemblage. It would perhaps be better in such cases to consider the classification as 'CG1+2', rather than CG1 and CG2, to remind ourselves that these are not independent groups, and that the actual balance of ubiquities between CG1 and CG2 may not be significant. In the second example, which results in a similar copper group classification, but is actually two independent distributions, profile analysis would immediately show that they were not from the same distribution. In this case the balance of ubiquities between CG1 and CG2 could then be significant.

One purpose of carrying out profile analysis of the alloying elements within specific Copper Groups is to see if there is any association between the type of copper being used (as defined by the Copper Groups) and the type of alloy being made from that copper. For example, if, in an assemblage of contemporary objects from a particular site, it turned out that copper of CG1 was always used to make unalloyed objects, whereas CG3 only appeared when associated with tin to make bronze objects, then we might postulate that the unalloyed objects containing CG1 came from one specific source, and that the combination of CG3 + Sn came from somewhere else. Moreover, we might also consider whether this indicated that the metal forming the combination of CG3 + Sn was imported to the site as finished objects, or as pre-alloyed ingots to be made into objects on site, or whether some other factor within the organisation of the local foundries resulted in the alloying of tin only with copper of type CG3, but not with CG1. A critical part of this analysis would also be to relate the typology of the objects to the two identified patterns of unalloyed CG1 and CG3 + Sn. As discussed in the next chapter, profile analysis of the tin associated with CG3 could also reveal whether this bronze was a primary (i.e., with a deliberately controlled composition) or a secondary alloy (with little or no control). This in turn could relate to how much metal recycling was being practiced. An example of this sort of analysis using Qing dynasty Chinese copper alloy coinage is given in the next Chapter.

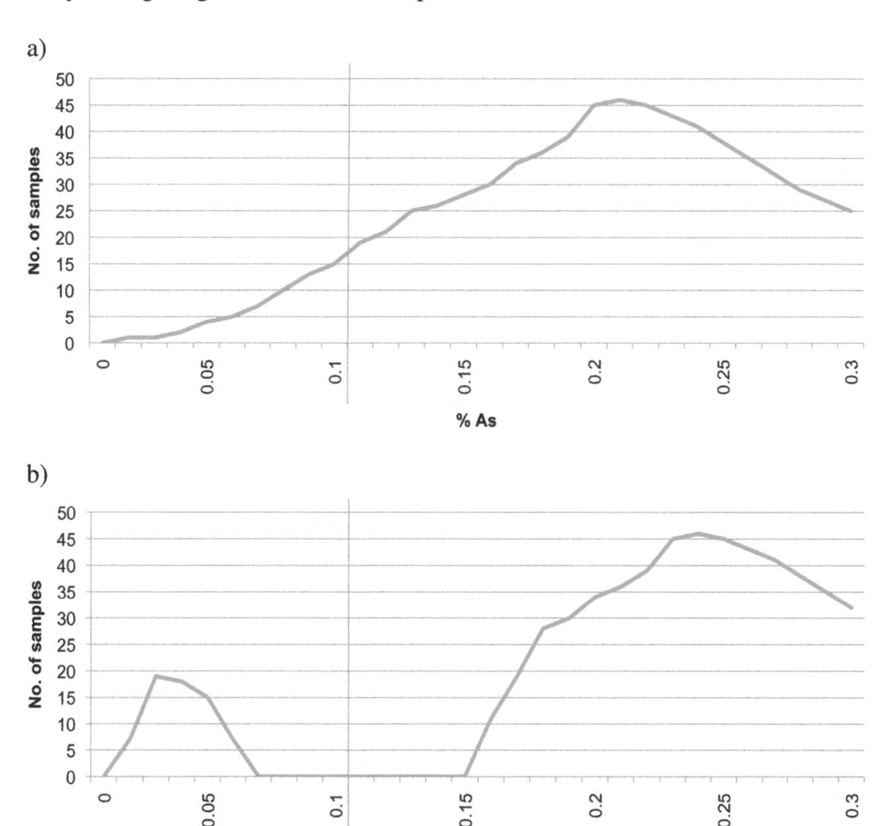

Figure 7:
Two simulated distributions of arsenic: a) a continuous distribution with some values below the cut-off at 0.1%, and b) two discrete distributions, one above and one below the cut-off at 0.1%

Suffice to say that it is the careful unpicking of the data within the Copper Groups (including the use of profile analysis and mapping) in combination with typological and archaeological evidence that can lead to significant insights, not only into the patterns of metal movement, but also possibly into the ways in which the metals are moving (i.e., as a specific pre-formed alloys, or as raw metal).

Mapping

One of the major ways of displaying Copper Group data is to map them. We use mapping to determine the extent, movement and timing of the circulation of particular metal (and alloy) types. The first maps we produced simply plotted by hand the ubiquity of each copper group across a map of Europe for a specified time period. In this case, ubiquity was calculated as the number of objects allocated to a particular copper group as a percentage of the total number of objects in a specific region (i.e., the assemblage in this case is defined as the totality of analysed objects in a specified region for a specified time period). Originally, for Europe, we used the regions as defined in the SAM project, since the majority of the data for Europe were derived from that project. An example is shown in **Figure 8**, which plots the distribution of CG2 metal (copper with only arsenic) for the Early Bronze Age 1 period, approximately 2000– 700 BCE (Bray 2009). Some of the objects contain tin, but, as described above, this is stripped out before allocating objects to Copper Groups.

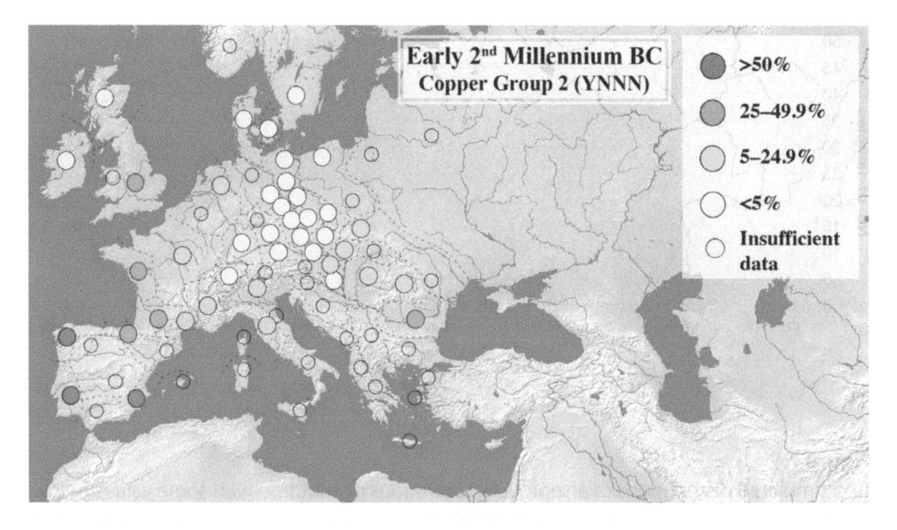

Figure 8:
Ubiquity of CG2 (Cu + As) across Europe in the Early Bronze Age (Bray and Pollard 2012).

The colour coding shows the frequency of occurrence of CG2 in the regional assemblage. It is highly regionalised, with the coast of Iberia showing very high ubiquities (>50%), whereas Germany has less than 5% of such metal. We concluded from this that a source of this type of copper is likely to be found in Iberia, and that there might be a coastal movement of this metal up the Atlantic coast of France and then into the UK, since there is a significant presence of such metal in these regions (25–50% ubiquity). From the archaeological data, we hypothesised that metal in the form of axes was being transported up the coast, but most likely in a series of short distance steps, and hence taking some time to move. The changing typology of the axes suggests also that they are being re-cast to meet local requirements of form at each stage of the journey. This hypothesis is supported by the declining average content of arsenic in the axes with increasing distance from Iberia, which is consistent with the model for arsenic loss on recycling, as described below. We should emphasise that this attribution of likely sources for Group 2 copper is obtained purely from the distribution of the compositions of the artefacts themselves, and is postulated without any reference to ore data or known Bronze Age mining sites. We can infer the existence of a source for a particular type of copper from ubiquity mapping—'hot spots' in the ubiquity map corresponding to likely source areas. This then allows us to cross-check against the locations of known mining or production sites and to compare ore geochemistry with that of the copper group, to identify the likely production sites within these broad areas.

Another possible source of CG2 metal in Europe is partially suggested in Figure 8 by the appearance of higher ubiquities of Group 2 metal in western Anatolia and the Balkans, but this particular region does not show the same high ubiquities as seen in Iberia, perhaps suggesting that the source of this metal lies further east, and that these points represent the 'tail' of a larger distribution centred off the map to the east. The distribution of this copper group throughout the Near East has been confirmed by subsequent work extending the map eastwards (Cuénod *et al*. 2015). **Figure 9** shows the original map inserted into a larger map, confirming the significant existence of Group 2 metal in Anatolia and western Asia, and suggesting possible source regions in Cyprus and Afghanistan. Current work (Howarth in prep.) will substantially add to this picture by adding more data from the Levant and the Caucasus. The combination of these two maps, however, demonstrate very clearly the advantage of using a common approach to the data (i.e., the principle of *universality*), in that the data from the second study focussing on Iran could be directly added to the original data from Europe.

The data in Figures 8 and 9 are still plotted on an arbitrary regional basis—in other words, the ubiquity is calculated as the proportion of objects classified as CG2 in a specified geographical area, which is often based on modern political geography. With the gradually increasing volume of data in our GIS database, we can now replace these arbitrary regions with a range of different possible areal units. These can include simple grid squares of equal size, or polygons which reflect the amount of data within them, or can be defined by geographical features such as watersheds (see Chapter 7). **Figure 10** shows the ubiquity of CG2 for the entire Bronze Age dataset across Eurasia based on a 1° WGS84 grid, confirming the importance of Group 2 sources in Iberia and western Asia, and also showing a strong presence along the Steppe belt, but with no significant occurrence in China.

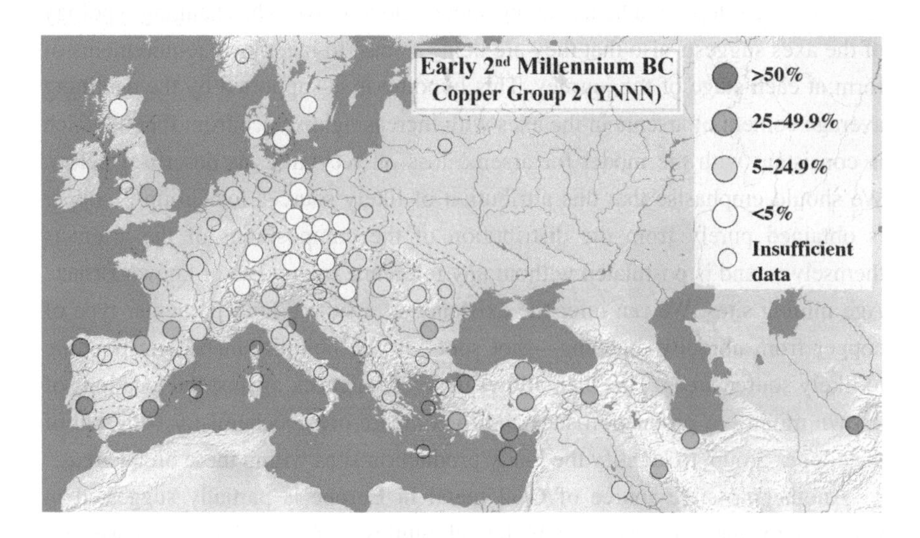

Figure 9:
Ubiquity of CG2 (Cu + As) across Europe and Western Asia in the early second millennium BCE (Bray *et al.* 2015).

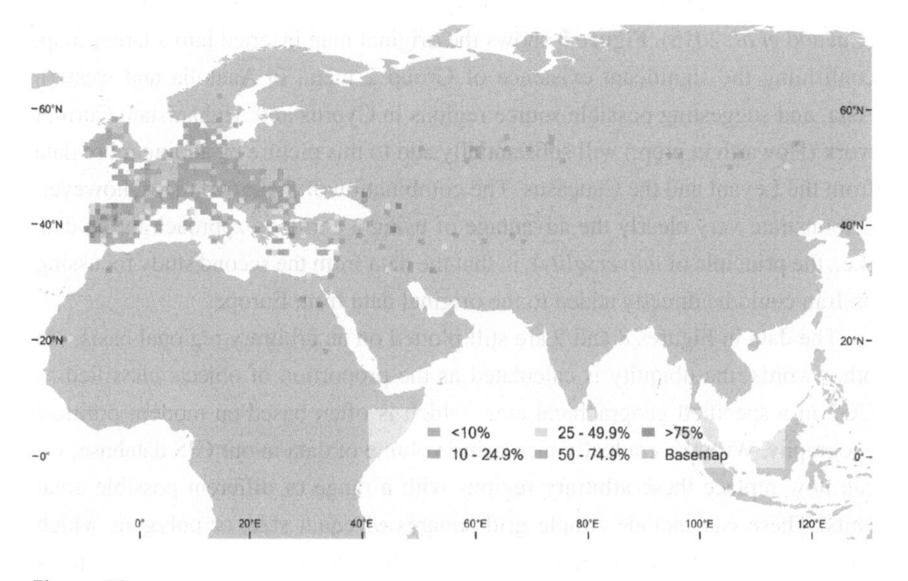

Figure 10:
Ubiquity of CG2 (Cu + As) across Eurasia (FLAME database).

Figure 10 shows that there are appear to be two or three centres of CG2 (Cu + As) use across Eurasia during the Bronze Age, focussed on i) the Iberian peninsula, ii) Anatolia, the Caucasus and southern Central Asia, and iii) the Altai mountains. Neither central and eastern Europe, nor China, have much significant presence of Group 2 copper. For regions where we actually have data, it is therefore obvious that copper is present, but not classified as CG2. **Figure 11** is a map of the ubiquity of CG16 (Cu + As, Sb, Ni, Ag) across Europe, to be compared with Figure 10. This shows an almost mirror image of the distribution of CG2—CG16 is virtually absent in Atlantic Europe and the eastern Mediterranean, but shows a strong presence in central Europe north of the Alps, with a focus on modern Germany, and a 'halo' of weakening ubiquity as the distance from Germany increases. This probably represents the output of the central German or Austrian fahlore sources, and the halo shows the extent of the distribution of such metal, diminishing in ubiquity as it competes with (or is mixed with) other types of copper. The value of such sets of maps is that they show the extent of movement of particular types of metal, and hence highlight the different traditions of metal use in different cultures.

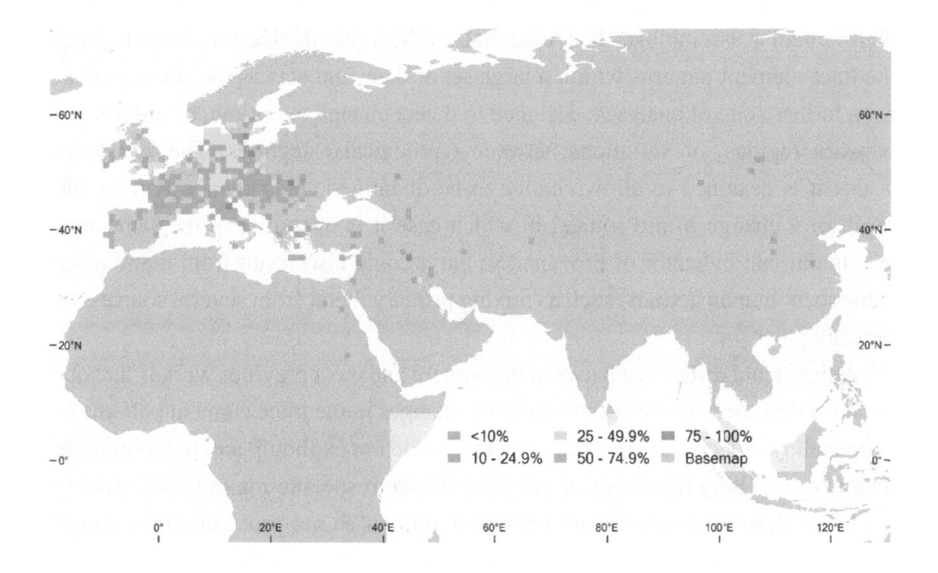

Figure 11:
Ubiquity of CG16 (Cu + As, Sb, Ni, Ag) across Eurasia, to be compared with Fig. 10 (FLAME database).

A key issue in this mapping is obviously the choice of chronological resolution and the definition of the size of each region. Often these decisions are imposed by the quality of the archaeological chronologies available for the metalwork, combined with the need to keep the numbers of objects in each regional and chronological group above a meaningful minimum. As noted above, the methodology becomes increasingly powerful when combined with Geographical Information Systems (GIS: e.g., Perucchetti *et al.* 2015, Peruchetti 2017: see Chapter 7), which, apart from providing more flexible mapping, also allows the use of geospatial statistics to investigate the relationship between metal distributions and transport routes, or topographical features such as mountains and river systems.

But what is a 'Copper Group'?

A key question to consider at this stage is 'what do we actually mean by the term Copper Group?' As previously noted, the concept of a Copper Group cannot be equated simply with ideas of provenance—objects within one Copper Group could come from more than one place, and one mine could produce copper of more than one group. The ubiquity of a particular copper group in space or time is simply an expression of how common a particular combination of trace elements is within that particular assemblage. It is essentially a heuristic device for characterising the trace element patterns within a large set of data, and acts as the starting point for a further suite of analyses, designed to detect change over time, or differences between regions, or variations between typological categories. The key point is that it is designed to allow *change* to be quantified. This change *may* be the result of a change in ore source, in which case it is operating in the same way as a traditional indicator of provenance, but it could also result from many other subsequent human actions, such as mixing primary metal from several sources, or recycling.

A significant difference between this system and most previous work is that our model is designed specifically to quantify changes in the trace element patterns in copper alloy *objects*, i.e., in worked metal. We do not explicitly aim to incorporate data from smelting debris, or of ore minerals from specific mining sites. That is not to say that these data are not important, but that in our procedure they do *not* form part of the initial mapping of object data. This is deliberate, and, we would argue, beneficial. As shown above, mapping particular Copper Groups as found in objects can lead to the identification of geographical 'hot spots', where a particular copper group represents the overwhelmingly dominant trace element pattern in that region. It is not unreasonable to assume that such 'hot spots' correspond to the output of a particular mine or mining region, whose ore mineralogy generally gives rise to a particular pattern of trace elements. The mapping also shows the extent of the circulation of such metal, and the fall-off in this circulation, as well as defining the likely location of the mine in broad terms. It is then a matter of turning to the published data on known Bronze Age mining (e.g., for Europe, O'Brien 2015), and comparing the ore mineralogy and/or the smelting products of that mine with the dominant copper group in that area. If the Copper Groups match the output of the mine, then we have two *independent* lines of evidence pointing to the exploitation of that mine in prehistory. Moreover, from the mapping we can estimate the extent of dispersal of the mining products, and also get some independent evidence of the date of exploitation of the mining region from the dating of the objects likely to have been made from the ouptut of that mine. In other words, we can achieve the goals of provenance studies from a completely different direction, and even add some additional useful spatial and temporal data.

Mixing and recycling

It is obvious that projects which focus exclusively on searching for a direct link between the trace elements (or isotopes) in metal objects and those in objects of known provenance or in mining debris (ores and waste metal) will produce increasingly diffuse results if metal or objects from different sources are mixed together. Many if not most previous authors have been aware of these problems, and several authors have cited mixing as a possible reason why the outcomes of provenance studies based on trace elements can become increasingly uninterpretable (e.g., Tylecote 1970). However, few if any have proposed a *practical* means of identifying and addressing this situation. The FLAME 'form and flow' model, focussing as it does on change in the flow of metal compositions, is deliberately designed to reveal such information.

One of the ways that we may be able to detect mixing and recycling is by observing systematic changes to the trace element distributions within and between assemblages over time. If we have two stocks of metal, one (A) containing copper with silver but without nickel (CG5), and the other (B) without silver but with nickel (CG4), then if these stocks are melted together in equal quantities, then we would expect the resulting metal to have roughly half the amount of silver as A and half the amount of nickel as B. Unequal quantities melted together would give *pro rata* results. Depending on how much silver and nickel are in the original metals, the result could be an assemblage of metal assigned to CG8 (Cu + Ag, Ni), and to one of CG4 and CG5, or both. Thus the appearance of a significant presence of CG8 in an assemblage related in some way to assemblages containing both CG4 and CG5 is suggestive of mixing between these two groups (**Table 4**).

	1	2	3	4	5	6	7	8	9	10	11	12	13	14	15	16
	clean	As	Sb	Ag	Ni	AsSb	SbAg	AgNi	AsAg	SbNi	AsNi	AsSbAg	SbAgNi	AsSbNi	AsAgNi	AsSbAgNi
Assemblage A	0	0	0	0	100	0	0	0	0	0	0	0	0	0	0	0
Assemblage B	0	0	0	100	0	0	0	0	0	0	0	0	0	0	0	0
A+B	0	0	0	x	y	0	0	100-(x+y)	0	0	0	0	0	0	0	0

Table 4:
Hypothetical mixing of Assemblage A (CG5) and Assemblage B (CG4), to give an assemblage containing CGs 8, 4 and 5, depending on the amount of silver and nickel in A and B, and the mixing ratio.

	1	2	3	4	5	6	7	8	9	10	11	12	13	14	15	16	
	Clean	CuAs	CuSb	CuAg	CuNi	CuAsSb	CuSbAg	CuAgNi	CuAsAg	CuSbNi	CuAsNi	CuAsSbAg	CuSbAgNi	CuAsSbNi	CuAsAgNi	CuAsSbAgNi	Samples
Roman	12.8	0.0	62.8	4.3	0.0	1.1	19.1	0.0	0.0	0.0	0.0	0.0	0.0	0.0	0.0	0.0	94
Early Saxon	3.7	0.3	28.9	5.0	0.3	2.1	48.3	0.0	0.5	1.1	1.3	4.0	0.5	1.6	0.3	2.1	377
Middle Saxon	8.7	8.7	14.1	2.0	2.7	7.4	34.2	0.7	2.0	0.0	1.3	14.1	0.0	1.3	0.7	2.0	149
Late Saxon	13.7	2.7	41.1	5.5	0.0	5.5	17.8	0.0	0.0	0.0	0.0	11.0	0.0	0.0	0.0	2.7	73
Early Medieval	13.8	1.7	25.9	1.7	1.7	27.6	12.1	0.0	0.0	0.0	1.7	5.2	0.0	0.0	0.0	8.6	58
Late Medieval	9.6	1.1	7.0	10.3	5.1	20.2	6.6	0.4	0.7	1.5	0.7	23.5	0.0	2.6	4.0	6.6	272
Post Medieval	9.9	2.4	4.7	8.0	24.1	1.9	4.7	1.9	0.5	1.9	5.2	13.2	0.9	2.8	10.4	7.5	212

Table 5:
Change of Copper Groups in English copper alloy objects from Roman to Post-Medieval times (table from Pollard ***et al.*** 2015; data from Blades 1995).

A possible example of this is shown in the data from Blades (1995), who published in his thesis more than 1200 high quality analyses of everyday English copper alloy objects from the end of the Roman period into the medieval era. **Table 5** shows the allocation of all of these data into Copper Group by period. The Roman assemblage is dominated by CG3 (Cu + Sb), with a contribution from CG7 (Cu + Sb, Ag). In Britain this combination of copper groups, both containing antimony with no arsenic, is uniquely Roman, and as yet we do not know where it came from. Possibly CG3 (Cu + Sb) represents partially refined copper originally containing both arsenic and antimony, but from which the arsenic has been removed. As discussed in Pollard *et al.* (2015), the major change in English copper supply came not at the end of Roman period (c. 400 CE), but at the onset of the Late Saxon period (c. 850 CE), several hundred years later. The drop in the ubiquity of CG7

(Cu + Sb, Ag) at the beginning of the Late Saxon period is accompanied by a rise in CG3 (Cu + Sb) during the Late Saxon period, followed by a rise in CG6 (Cu + As, Sb) and CG12 (Cu + As, Sb, Ag) in the Early Medieval period (11th – 13th C. CE). The rise in CG12 might reflect a new source of metal, but could also include some mixing between newly arrived CG6 with already circulating (but declining) CG7, which if mixed together would provide arsenic, antimony and silver to create CG12. This combination is more likely than mixing of CGs 2 (As), 3 (Sb) or 6 (As+Sb) with a silver-containing copper, since these are virtually absent in the circulation.

Such simple linear mixing models rapidly become complicated, depending on the distribution of each element in the assemblages, and on the 'mixing recipe'. Nevertheless, the consequences are in principle calculable. This is only true, however, if there are no losses of the trace elements due to either chemical or physical processes during the melting process, or subsequent working by casting or hammering. This raises the important question of the stability of various trace elements during melting and processing. The temperature required to melt copper alloys is in most cases likely to be in excess of 1000^{0} C. At these temperatures, we may expect to lose the more volatile trace elements, either through direct volatalization of the metal, or through oxidation and then volatalization. The first aspect of this behaviour can be broadly predicted from the Ellingham diagram (**Figure 12**). This plots, amongst other information, the affinity of each metal for oxygen at different temperatures, and allows predictions to be made about which metal in a mixture is most likely to oxidize first at a particular temperature and oxygen availability. Like all chemical equilibrium models (for such it is), it is only a partial guide as to what might happen in the real world, but it is the foundation of modern chemical metallurgy (Beeley 2001). The precise rates of oxidation (i.e., the kinetics of the situation) are affected by many factors, such as the degree of agitation of the molten metal, mutual solubilities between pairs of metals, thermodynamic activities (i.e., how the presence of one metal affects the behaviour of another), as well as the energies of volatilization and the partial pressures of oxygen and other gases (Merkel 1982, 30; Beeley 2001, 31). The importance of such factors is shown by the fact that they are exploited by the modern copper industry in order to oxygen refine the raw copper produced by the smelter (Copper Development Association 2011), and also that such selective oxidations are now an important part of the metal recycling industry, e.g., recovering copper from mobile phones (Kaspar *et al.* 2011). Consequently, there is an extensive literature on oxidation effects in molten metals that is useful for considering in the archaeological context: a copper alloy object will be depleted in certain of its metals depending on how many times it has been reheated, and to what temperature (e.g., Hampton *et al.*

1965; Charles 1980; Pickles 1998; Beeley 2001: 497; Tanahashi *et al.* 2005; Lee *et al.* 2009). Unfortunately, however, the modern metal recycling industry is not interested in the same combinations of trace elements as we find in Bronze Age alloys, so a certain amount of experimental work has been necessary to understand these systems (Sabatini 2017).

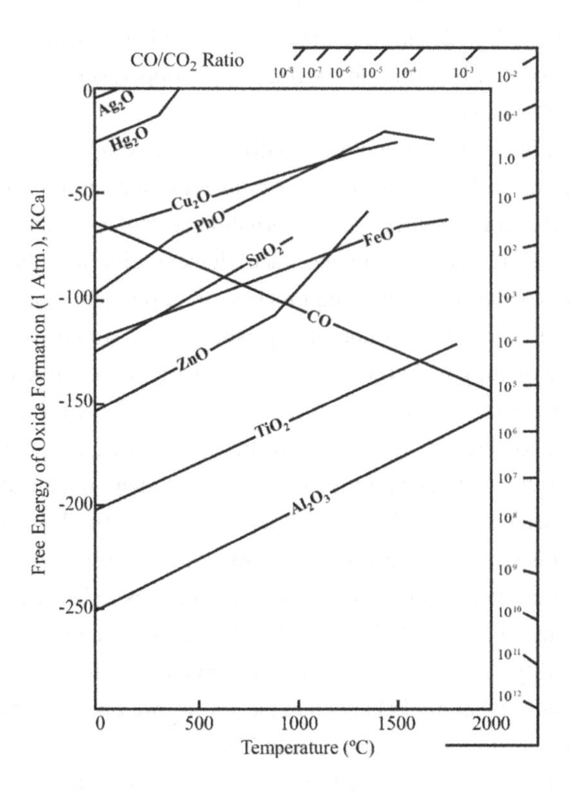

Figure 12:
The Ellingham diagram (Bray and Pollard (2012), based on Killick (2001) and Anderson (1930)).

Considering these observations based on modern data from the metal recycling industry, it should come as no great surprise that, under certain conditions, melting impure copper can result in the partial loss of certain trace elements through oxidation, particularly arsenic and antimony. This was convincingly shown for arsenic by McKerrell and Tylecote (1972), who demonstrated that under reducing conditions (nitrogen atmosphere), no arsenic was lost, but under oxidising (atmospheric) conditions it was lost rapidly (**Figure 13**). In the same paper, they carried out one experiment on the loss of antimony from a 4.3% antimonial copper (their table 1), and showed that the loss of antimony was comparable to that of arsenic under approximately the same conditions. Other workers have elaborated on these observations (e.g., Merkel 1982; Pernicka 1999; Earl and Adriaens 2000), but, in short, we may assume that arsenic and antimony are relatively vulnerable to oxidation and loss on recycling if the metal goes through a melting process with sufficient oxygen availability, whilst silver and nickel are much more resistant to removal. Hence, because of their lower vulnerability to oxidative loss, a reduction in the average silver or nickel content within an assemblage is more likely to be caused by dilution with a unit of silver- or nickel-poor metal rather than a consequence of oxidative losses. Summarising the causes of chemical variation in smelted copper, Pernicka (1999) concluded that a very limited range of trace elements are invulnerable to oxidative losses, and can therefore be directly related to the provenance of the ore (most significantly gold (Au), silver (Ag), bismuth (Bi), iridium (Ir) and nickel (Ni)), whilst a large number of the others are prone to loss (including tin (Sn), providing it is present at less than c. 1%, zinc (Zn), lead (Pb) if less than 5%, plus arsenic (As), cadmium (Cd), cobalt (Co), indium (In) and mercury (Hg)) and, according to Pernicka, are therefore likely to be indicators of either technology or provenance. Although his study is focussed on the suitability of particular elements in copper for conventional provenance studies, it does show very clearly that oxidative losses do occur when copper is melted. Merkel (1982, 287) refers to the "confusion" that such differential oxidative losses could cause. However, if we shift our theoretical focus somewhat, we see that, rather than being a perturbing factor in the quest for provenance, these potential chemical changes under melting present us with an extremely powerful interpretative tool for understanding the interaction between humans and metal.

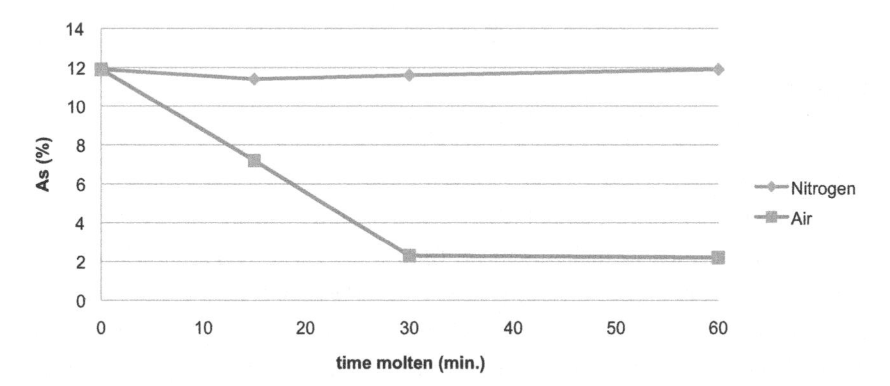

Figure 13:
Comparison of loss of arsenic from 12% arsenical bronze under oxidizing and reducing conditions (drawn from data in McKerrell and Tylecote 1972, Table 3)

It is clear that the rate of loss of the more vulnerable trace elements will depend on many factors apart from time, temperature and redox conditions. These include the way that the processes (melting and hot working) are carried out, and factors such as the influence of other elements in the metal (Sabatini 2017). In an attempt to construct a simple model to predict the rates of loss of arsenic and antimony from a copper melt, we combined the data of McKerrel and Tylecote (1972) for arsenic with observational data generated by Bray (2009) on the relative rates of loss of arsenic and antimony in objects from Ross Island, and calculated exponential parameters for the rates of loss for arsenic and antimony under oxidizing conditions. This model (**Figure 14**) simply assumes that the rate of loss of an element is proportional to the amount of that element left at time t.

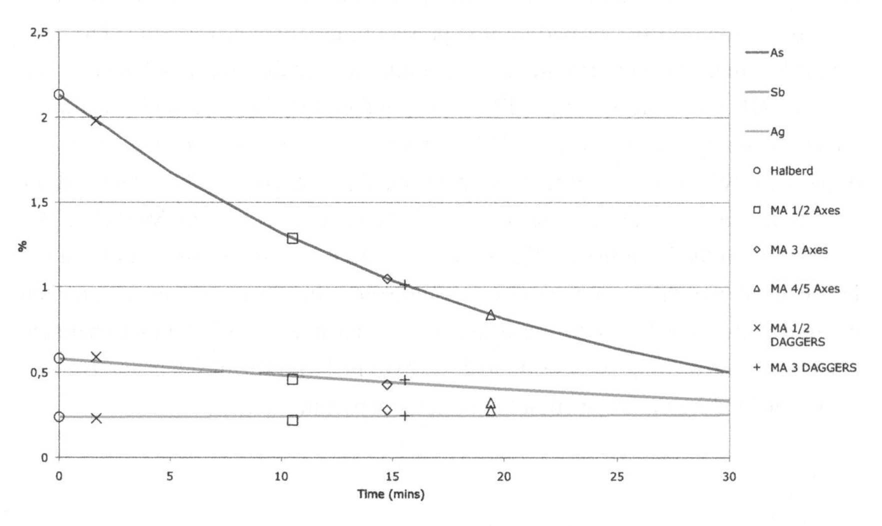

Figure 14:
Theoretical calculation of the rate of loss of arsenic and antimony from Ross Island metal.

The starting point for this model (Cu + 2.2% As, 0.6% Sb and 0.25% Ag) is the average composition of halberds made from Ross Island metal (Ireland) in the British and Irish Early Bronze Age (c. 2500 – 1700 BCE), since these objects have the highest average levels of arsenic and antimony of any objects attributed to the Ross Island source (Bray 2009). Applying the calculated rate constants (-0.05 for arsenic and -0.02 for antimony, and a linear evaporative loss for copper) shows the expected rapid loss of arsenic and a slower loss of antimony, with silver slightly increasing as a result of the loss of the other elements in the melt, primarily copper (**Figure 14**). Superimposed on this figure are the average compositions of other classes of artefacts attributed to Ross Island metal, but from different periods, namely axes from Metal Ages (MA) 1/2, 3, 4/5, and daggers from MA 1/2 and 3. Metal Ages are successive chronological periods during the British and Irish Bronze Age, defined by metal hoards from specific type-sites—see Needham (1996). The position of these other groups on the graph is unknown, and is simply fitted by locating the average value of %As along the exponential decay curve for arsenic. Not unexpectedly, perhaps, given the way the model was constructed, the associated values of antimony and silver fit well on their respective calculated lines. If this model is at all realistic, however, the horizontal axis indicates the total average length of time each category of object was exposed to high temperature oxidative conditions, either being molten, or subject to hot working. These predicted times range from around 2 minutes for MA 1/2 daggers, up to around 17 minutes for MA 4/5 axes. Noticeably, the times predicted for the earlier objects (MA 1/2) are shorter than the later objects, consistent with the idea that later objects are made from recycled earlier objects and have therefore 'accumulated' more time at high temperatures, resulting in greater oxidative losses. The difference between the composition of axes and daggers in MA 1/2 is an important but not new observation. Traditionally it would be explained by selection of ore or metal for the specific purpose of producing either axes or daggers, despite the fact that the visual differences are probably too subtle to have been observable by Bronze Age metalworkers, although arguably the working properties might have been slightly different. The model presented here suggests an alternative explanation—the starting material was broadly the same, but the manufacturing process of halberds, axes and daggers differed, resulting in axes being exposed to longer periods at high temperatures. This could be a difference between open and closed moulds, but could equally relate to the time spent on hot working. This case study is probably a rare but not unique chance to study such effects, since it comes from a time and place when there was only one source of primary metal, and it represents a relatively isolated system.

The model presented here is also consistent with the discussion in another section of the paper by McKerrell and Tylecote (1972, 213), where they consider the different compositions of halberds and the rivets associated with them. For practical reasons, the rivets need to be softer than the metal they hold together, so that they can be hammered into shape without damaging the hilt. McKerrell and Tylecote noticed that in a set of 12 Scottish and Irish halberds, the rivets are, with one exception, always lower in arsenic and antimony than the host object, but also that the rivets as a group do not have a standard composition. This, they suggest, is deliberate, and a consequence of higher volatalization of the trace elements when the same material that is used to make the halberd is further worked to make the rivets. Thus there is no need to invoke specialized selection of material for halberds and rivets—it is a consequence of the process used to make the rivets, and is a deliberate and beneficial outcome.

One of the strengths of this model is that it can be used to explain the composition of groups of copper alloy objects that are either difficult to explain or are not currently linked to any known ore sources. Essentially our argument is that some groups of metal (those with the 'long branched complex biographies' described in Chapter 2) are so mixed and recycled that they no longer carry the characteristics of the mine from whence they originally came. An example is a group of 66 Early Bronze Age British and Irish copper-alloy artefacts that have a strange chemical signature of mostly copper and tin, but also with low but significant levels of silver (between 0.1 and 1%), with all other elements absent (below 0.1%). In Ireland, this composition makes up a significant proportion of the assemblage in the second half of the EBA (c. 2000 – 1500 BCE), comprising around 10% of all flanged axes, for example (Bray 2009). Under the simple paradigm of connecting chemistry to provenance, this chemical group is difficult to interpret, as there are no clear chemical links to a possible source of copper ore. Allowing for shifts in the chemical composition as a result of post-smelt alteration, however, gives this group an obvious identity as being heavily recycled Ross Island metal. The more vulnerable arsenic and antimony present in the fresh metal have been oxidised away and only the more noble silver remains, as predicted by the Ellingham diagram and experimental results (Figures 12 and 14). The increasing frequency of this depleted 'silver only' copper type towards the end of the EBA supports the possibility of its Ross Island origins. The mine closed around 2000 BCE, and old metal was therefore increasingly recycled over the next five centuries to leave objects with this strange composition. If this interpretation is correct, then it not only explains the origin of this unusual group of copper objects, but it also suggests that, at least in this particular case, metal was being recycled for up to 500 years after its original extraction.

The fact that we do not consider the levels of trace elements to be fixed within the metal flow is one of the major differences between our model and those that have gone before. It is for this reason that we refer to it as a 'dynamic' rather than a 'static' model. We are not, of course, suggesting that in the normal course of events the trace element composition of a single specific object is going to change, although of course a different process—selective corrosion—can and does cause such changes. The changes we are talking about can only occur when we consider the flow model, whereby the composition of the *metal* flow can be affected by mixing different stocks of metal, or in some cases by recycling and re-melting material. Nor is it realistic to think of such a model in terms of single objects. As emphasized in Chapter 2, the model relates to assemblage properties rather than individual properties.

It has been long appreciated that certain trace elements are volatile in copper under hot oxidizing conditions, and the work of McKerrell and Tylecote (1972) amongst others has suggested that these properties were known and exploited by ancient metalworkers. We have taken this further by suggesting that these losses can be used to understand more about the biography of metal, so that, rather than being seen as a cause for confusion in the quest for provenance, such losses are a vital element in understanding the relationship between humans and metal. The model presented here is crude in the extreme, and needs better data from controlled laboratory experiments to generate the appropriate rate constants for specific circumstances. However, the concept of incorporating such ideas into a broader picture of the metal flow seems to be able to explain some observations without the need to resort to unknown ore sources, or the selection of special metals for specific purposes.

Specific issues when considering Copper Groups

Silver in lead

The theory described above makes the initial assumption that the four trace elements considered (As, Sb, Ag, Ni) enter the copper alloy *only* through being associated with the copper. Another way that silver can enter the alloy is by being associated with the lead rather than the copper, especially if the object contains added lead. Argentiferous galena (PbS), containing several hundred ppm of silver, has always been one of the richest sources of silver, and therefore any lead smelted from such an ore, if not de-silvered, could potentially bring significant amounts of silver into the object. This can easily be tested, by plotting silver vs lead in a particular assemblage. If there is a strong positive correlation, and the ratio of silver to lead is within the known range of concentrations of silver in argentiferous galena, then it is highly likely that the silver is associated with the lead, and not the copper. This is clearly an important observation, but, from the point of view of view of creating Copper Groups, it does not change the process, because we see the use of Copper Groups as being primarily a means of detecting change in the composition in the metal flow. It would, of course, be a mistake to seek an argentiferous copper source on the basis of the presence of silver in the alloy, if it had actually entered with the lead. Again, because our method does not make the assumption that the silver characterizes the copper source, we can simply assume that the presence of silver is yet another marker in the life history of the flow of metal. Rather than being a problem, the presence of silver from lead is another piece of information we can use to our advantage.

Arsenic—trace element or alloying element?

Effectively the system described here treats arsenic (As) as a trace element, most likely entering the metal as a contaminant from the source of copper ore. However, it is widely accepted that arsenic was probably used as an alloying element, especially in the earlier stages of the western Eurasian Bronze Age, (e.g., Budd and Ottaway 1990) before tin became the major alloying element. It is also possible that antimony was used, but less frequently, as a deliberate alloying agent, especially in the Caucasus. Objects made from pure antimony have been reported from there, giving rise to the likelihood that metallic antimony was available for direct alloying. No metallic arsenic has ever been identified in the ancient world, and is not very likely because although arsenic does occur native, producing metallic arsenic from sulfide ores requires roasting without air. It is more likely that any deliberate alloying of copper with arsenic would have necessitated co-smelting copper and arsenical ores, or perhaps the use of speiss—an iron-arsenic alloy deliberately produced from arsenopyrite—as an alternative means of adding arsenic to copper alloys: see Rehren *et al.* (2012). It is therefore likely that, given the volatility of arsenic, deliberate production of arsenical copper by co-smelting would have resulted in very variable alloy compositions, from more than 10% arsenic down to less than 1%.

There are two points to be made here. One is that using our system it does not affect the interpretative process if arsenic is deliberately added (i.e., it is an alloying element) or is a trace element (assumed to be unintentional), providing we do not use the presence of arsenic as an indicator of copper ore source when it has in fact been deliberately added. Since our aim is primarily to characterize changes in the nature of the metal flow rather than to specifically track a metal back to an ore source, treating arsenic as a trace contaminant when it is deliberately added is largely irrelevant in the preliminary stages of analysis. Secondly, as described above and in the next chapter, the classification into presence/absence is only the first stage of the analysis of either trace elements or alloying elements, and the second stage is usually to plot the elemental profile of that element within a particular assemblage. This is described in more detail in the next chapters, but suffice to say that with arsenic (or antimony) it is often possible to say whether we are dealing with an assemblage containing deliberate additions of arsenic, or with one dominated by either an accidental inclusion of arsenic, or heavy recycling.

Chapter 5

Alloying Elements and 'Alloy Types'

We take a similar approach to classifying archaeological copper alloys as we do to copper groups—a preliminary classification step based on presence/absence, this time of tin, lead and, if present, zinc, followed by ubiquity analysis, profiling and mapping. Traditionally, archaeological copper alloy types are classified using essentially the same definitions as used for modern alloys. These definitions (which are effectively production specifications) have emerged over several centuries as being optimum compositions for particular applications—machining alloys, casting alloys, corrosion resistant alloys, etc., and therefore represent compositions based on technological experience (**Figure 1**). They have the advantage that such definitions are widely understood, but we believe that they can hide important information when applied to archaeological objects. The use of these modern alloy definitions to describe archaeological objects implies that their alloy composition was deliberate and targeted at producing alloys with specifications that approximately correspond to modern definitions. This was clearly the case in some circumstances, but not in all. The chemical analyses of many archaeological objects do not correspond to simple modern definitions of alloys. This could mean that technological considerations were not the primary driver in selecting alloy formulations, or it could be that compositional control was not always exercised. For example, in modern terminology, an object is only called a 'brass' if it contains more than 8% zinc, or a 'bronze' if it has greater than 3% tin—in other words, if the level of the alloying element exceeds a value that is considered to have an effect on the visual or physical properties of the object and could therefore be interpreted as a conscious addition. Many archaeological artefacts contain alloying elements below such levels, and yet higher than would be expected if the sole source of such elements were contaminants in the copper ore sources. There has been much useful work recently relating the colour of the metal to the composition (e.g., Hosler 1995, Kuijpers 2013, Mödlinger *et al.* 2017, Radivojević *et al.* 2018), which has important implications for understanding the question of intentionality. Although we do not dispute for a moment that certain alloyed metals in antiquity *were* designed so as to achieve particular physical properties or a desired appearance of the finished product, it is the information contained in the 'non-standard' alloys that we particularly wish to capture by using a different approach.

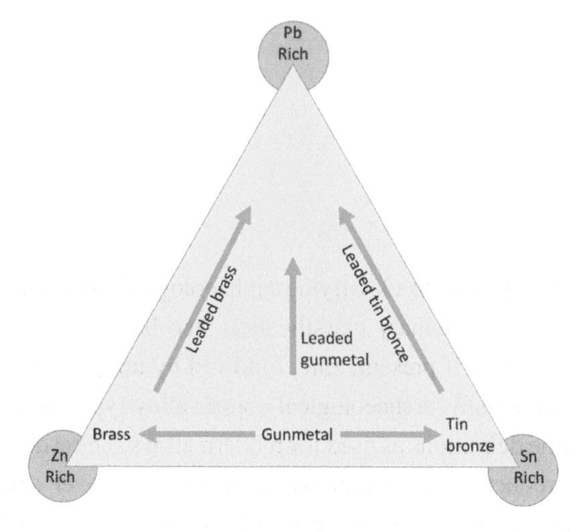

Figure 1:
Schematic diagram of the traditional classification system for modern copper alloys with lead, tin and zinc.

We have therefore adopted a presence/absence definition, in which we generally set the threshold of 'presence' at 1% for the alloying metals. Our definitions are shown in **Table 1**. Thus we would classify an alloy containing 92% Cu, 2% Sn, 2% Zn and 4% Pb as a 'leaded gunmetal' (LG), despite the fact that this composition does not correspond to the modern definition of a 'leaded gunmetal', which would typically have 5% each of tin, lead and zinc. Perhaps more controversially, we would classify an alloy consisting of 98.5% Cu with 1.5% Sn as a 'bronze' (B), or with 1.5% Zn as a 'brass' (BR), although neither conforms to such an alloy in modern terms (a modern bronze typically has 12% Sn, and a brass usually has more than 20% Zn). Nevertheless, we feel justified in doing so because such an approach tends to highlight the intermediate levels of alloying metals (above 'accidental' but below 'designed') which we think contain considerable information about the biography of the metal flow. Our argument is that if a copper alloy contains, say, 2% Sn, then this is unlikely to be the target level of Sn (since it has little effect on the colour or properties), nor is it *likely* to be an ore contaminant, although, particularly in the earliest phases of the metal age, it could be a consequence of co-smelting specific ores (e.g., Radivojević *et al.* (2013). Nonetheless it signifies the presence of tin in the metal flow. It could signify the mixing of a tin bronze (with more than c. 8% Sn) with copper metal containing little or no tin, or possibly just tin bronze that has been heavily recycled, since it has been shown that tin is lost on recycling (Godfrey 1996). There is thus considerable archaeological significance to this low level of tin, which would not be so easily recorded if modern definitions were used. It is nevertheless always important to remember the differences between our definitions and modern terminology.

Code	Name	Chemical Definition
C	Copper	Pb, Sn, Zn all <1%
LC	Leaded Copper	Pb > 1%; Sn, Zn both <1%
B	Bronze	Sn >1%; Pb, Zn both <1%
LB	Leaded Bronze	Sn, Pb both >1%; Zn <1.%
BR	Brass	Zn >1%; Sn, Pb both <1%
LBR	Leaded Brass	Zn, Pb both >1%; Sn <1%
G	Gunmetal	Zn, Sn both >1%; Pb<1%
LG	Leaded Gunmetal	Pb, Sn, Zn all >1%

Table 1:
Definition of alloy types used in this work.

This system has the property of highlighting rather than hiding the presence of such mixed or non-standard alloys within assemblages. If a high proportion of an assemblage is made up of such alloys, it suggests the possibility that those objects may be the result of mixing metals of more than one alloy type, rather than of deliberate alloy design. Rather than denying the existence of 'designed alloys', however, our methodology enables us to identify them clearly when they do appear in the metal flow. For example, in a study of first millennium CE copper alloys in Britain using these definitions (Pollard *et al.* 2015), we clearly see the arrival of brass as a designed alloy into Roman Britain in the Late Iron Age and 1st century CE, followed by a steady decline in its ubiquity from the second century onwards (**Figure 2a**). Furthermore, we see continuity of metal circulation from the first century through the Late Roman period (c. 400 CE) and into the Early Anglo-Saxon period, with a marked change occurring only in the Middle Saxon period (c. 750 CE), which we attribute to the arrival of fresh stocks of metal from northern Europe. Moreover, using the ubiquity of the quaternary alloy leaded gunmetal (LG as defined as above) as a proxy for the amount of recycled metal in circulation, we have suggested that by the end of the Early Saxon period, at least 70% of the analysed objects contained recycled Roman metal, suggesting a period of re-use of at least 250 years.

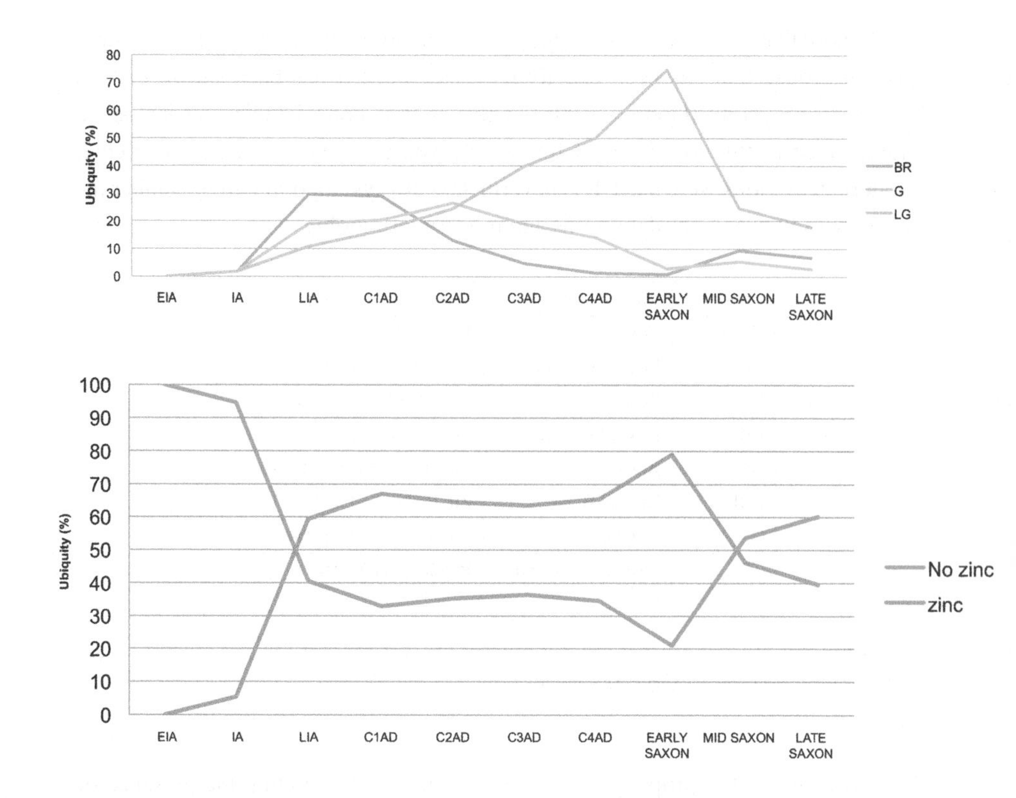

Figure 2:
Changes in the ubiquity of BR, G and LG over time in Roman Britain (Pollard ***et al.*** 2015).

Profiling alloy compositions

We can, however, take this methodology further, by considering the distribution of the concentration of alloying elements within a particular assemblage, and also by mapping the ubiquities of different types of alloys. For example, we can infer whether the bronzes in a certain region are primary alloys or contain recycled metal by considering the shape of the distribution of the tin concentrations within the assemblage. By *primary*, we mean that they are made to a specific (and therefore designed) composition, and do not contain significant amounts of recycled metal. Thus a unimodal approximately normal distribution of tin centred around a specific value (which is typically between 10 and 20%, depending on the quality or function of the alloy) indicates that the assemblage is likely to be made of a primary alloy (e.g., Cuénod *et al.* 2015; Hsu *et al.* 2016). In contrast, 'secondary' alloys tend to have a wider spread of values, often with a distribution which tends to increase as it approaches zero. This is potentially indicative of a recycled alloy, where copper and bronzes of different tin levels have been mixed together. These two situations are shown in **Figure 3**. The metal objects from Fu Hao's tomb in Anyang (the only unrobbed royal Shang tomb so far discovered) show a highly symmetric distribution of tin centred on 15–18%, indicative of a set of objects made from a primary alloy, whereas those from multiple tombs in the Western Area of Anyang show a preponderance of objects with low tin (<6–9%), with some potentially primary objects showing as peaks at higher tin values (Li 1982, Li *et al.* 1984). The implication here is that some, at least, of these objects have lower tin contents, perhaps as a result of recycling or dilution, indicating secondary alloys. Further discussion of comparing the profiles of alloying elements in different assemblages is discussed below.

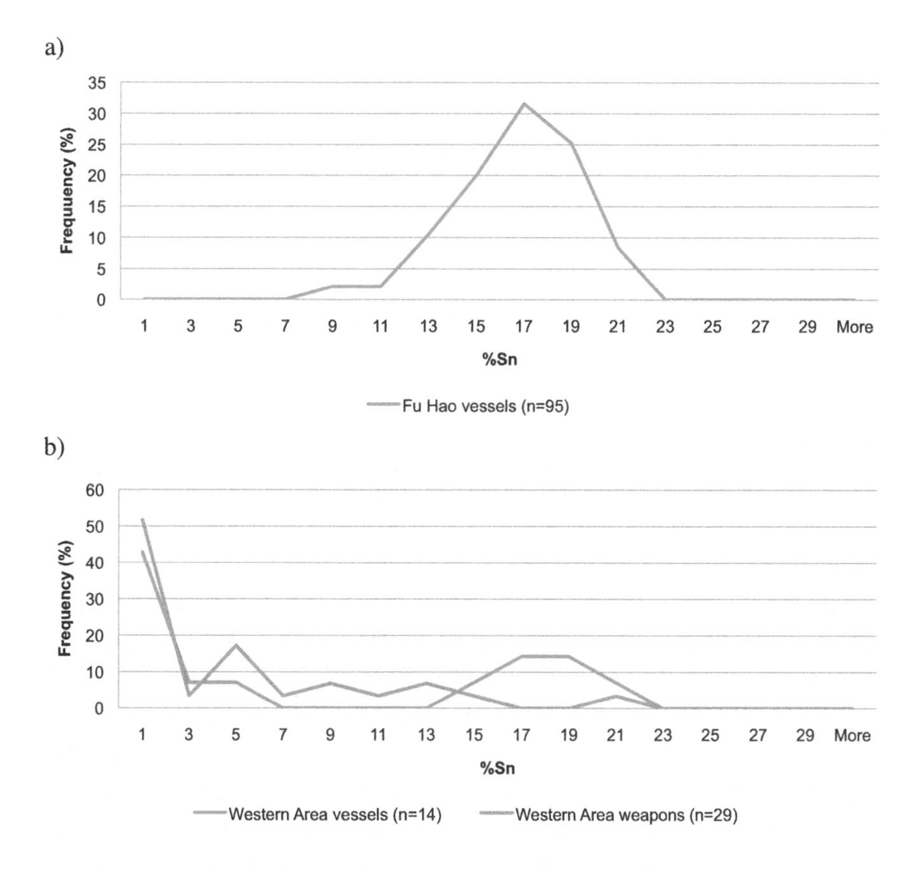

Figure 3:
Comparison of the tin profiles of a) an assemblage consisting of 'primary' alloy (from Fu Hao's tomb, Anyang) and b) an assemblage of secondary alloys (from the Western Area of Anyang) (Liu 2016, Fig. 211, based on data from Li (1982) and Li ***et al.*** (1984))

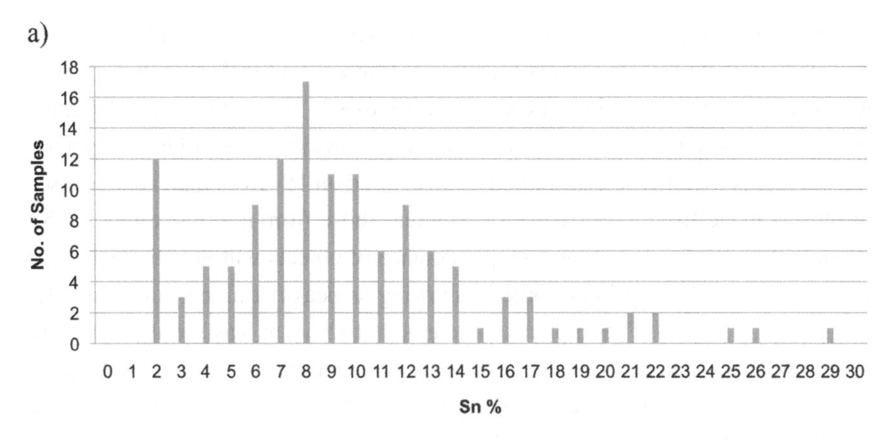

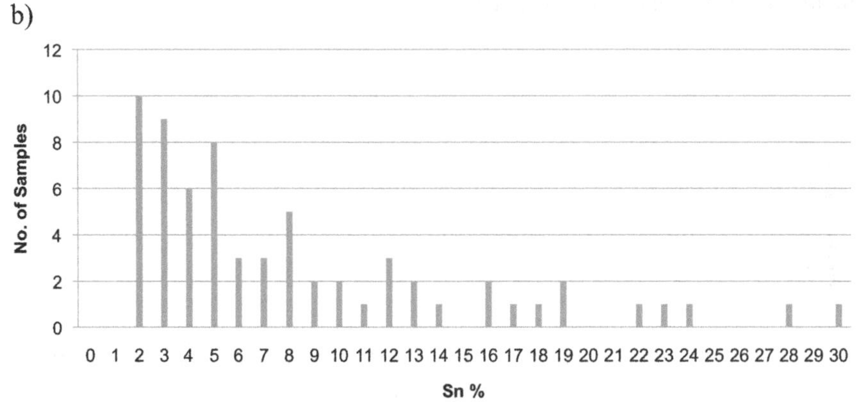

Figure 4:
Tin distribution in bronze objects assigned to a) CG2 and b) CG14 for Period 1 (early third to mid-second millennium inclusive) in Mesopotamia (Cuénod *et al.* 2015).

The utility of profiling the distribution of alloying metals (most commonly tin) within an assemblage can be further demonstrated by combining this profiling with the allocation of the objects to copper groups. This can reveal specific patterns of behaviour within the metal flow. For example, Cuénod (2013) used the analyses of around 5000 metal objects from the Bronze Age and Iron Age of Iran and the Near East to study the way in which tin was used and circulated in this region. Two profiles of tin in the bronzes from Mesopotamia dating from the early third to the mid-second millennia BCE (Period 1 in her terminology) were produced —one for objects allocated to CG2 (Cu + As), and one for CG14 (Cu + As, Sb, Ni) (**Figure 4**). Two very different profiles can be seen: for CG2 (Fig. 4a), the tin content presents a nearly symmetrical unimodal distribution with a mode around

8–10%. For CG14 (Fig 4b), the tin contents show a distribution skewed towards much lower values (2 to 5% Sn). The primary distribution of tin in the CG2 objects from Mesopotamia suggests that there was little mixing, re-melting, or secondary alloying occurring after the alloying process, as this would result in a wider and less regular range of tin compositions. Since Mesopotamia did not produce tin, the addition of tin at a late stage in the life of this copper group could either mean that Mesopotamia imported copper and tin separately and alloyed the metal once there, or that they imported a fairly standard pre-alloyed bronze and did not alter it after importation. In contrast, the tin content of objects made using copper of CG14 shows a typical 'secondary' profile, suggesting that these objects were made with less control of the tin content, which might be indicative of recycling. We can imagine that this copper was not the subject of an organised trade but reached Mesopotamia as copper and pre-alloyed bronze objects, perhaps as down-the-line trade or in events of booty-taking, and was subsequently treated as scrap material and recast to suit the Mesopotamian taste.

Mapping alloys

Figure 5 shows a distribution map of the tin concentration in bronze assemblages from Azerbaijan, Amlash, Luristan and Mesopotamia between the late second and mid-first millennium BCE (Cuénod *et al.* 2015).

The Mesopotamian tin profile is similar to that shown for CG14 in the earlier period in Figure 4, suggesting recycling and an overall paucity of tin. It contrasts greatly with the other three profiles from western Iran (Luristan and Amlash) and Azerbaijan, which show a primary signal, suggesting that tin was easily available in these areas. It is tempting to infer the presence of one or more tin sources in western Iran or eastern Anatolia from these observations. The existence of the tin mine at Deh Hosein in Luristan provides a potential source for the presence of tin, at least in Luristan (Nezafati *et al.* 2009).

Mapping the types of alloys used in specific regions can show large-scale patterns of metal circulation, as was done by Hsu (2016) in his study of the flows of copper and copper alloys in the Early Iron Age societies of the eastern Eurasian steppe and northern China. **Figure 6** shows the ubiquity of different classes of alloy in different regions across southern Siberia, Mongolia and into China. The size of each circle reflects the number of analyses within each assemblage.

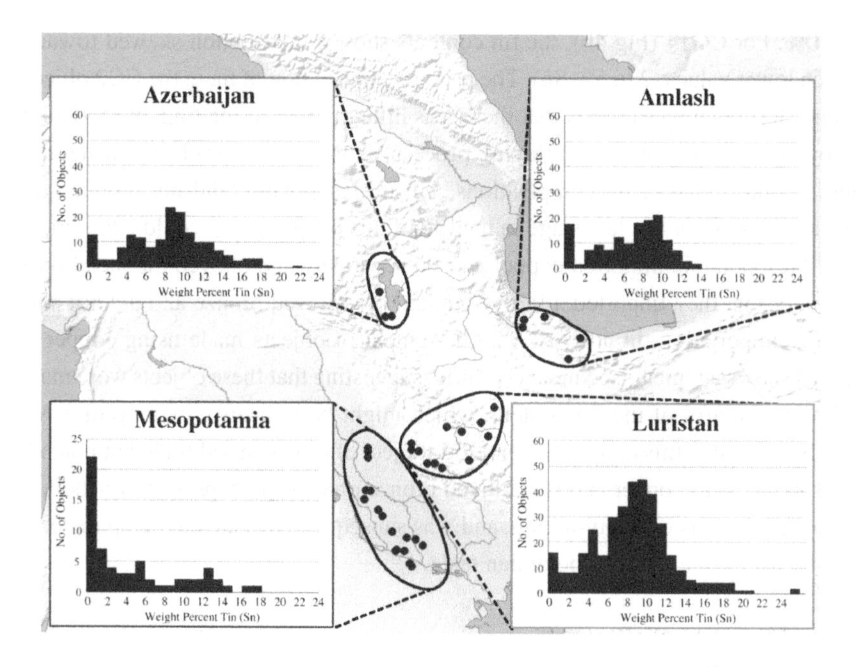

Figure 5:
Distribution of the tin concentration in assemblages from Azerbaijan, Amlash, Luristan and Mesopotamia between the late second and mid-first millennium BCE (Bray *et al.* (2015), redrawn from Cuénod *et al.* (2015)).

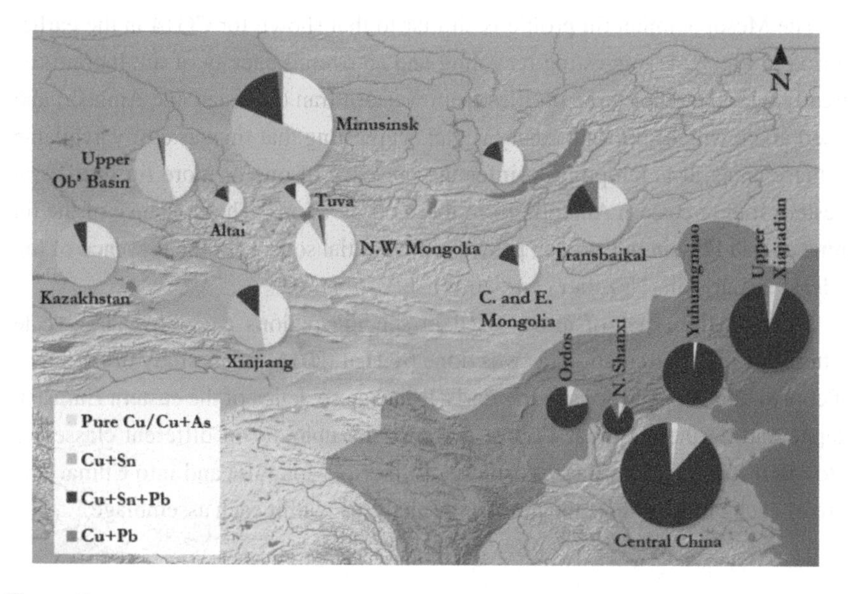

Figure 6:
Alloy Types in East Asia – 1st Millennium BCE (Hsu 2016).

There is clearly a distinctive difference between the alloys used in Siberia and Mongolia and those in Central China. The former assemblages are dominated by either unalloyed copper, copper-arsenic alloys (see previous chapter for a discussion of Cu-As alloys), or tin bronzes, whereas Central China is almost exclusively made up of the ternary alloy leaded bronze (Cu + Sn + Pb). The interesting part of this diagram is the area shaded in dark grey—a region known as the 'Arc', which effectively forms the border between Central China and the Steppe (Tong 1987). The alloys used in this region are also almost exclusively leaded bronze, corresponding to the tradition in Central China rather than the Steppe. The metal objects found here, however, are almost exclusively 'Steppe-style'—knives, horse fittings and items of personal adornment—raising the question of why such Steppe style objects in this border region are made to an alloy recipe from China rather than the Steppe. Since a few objects of unquestionable Central Plain origin are also found here, it is possible that metal was imported, looted or gifted from China, and re-made into Steppe-style artefacts. The fact that the Chinese objects are mostly ritual bronzes designed for a ritual which was not shared by Steppe people might support the hypothesis that such objects were re-purposed by making new objects on arrival in the Arc (Rawson 2015). An alternative possibility is that the concept of alloying was adopted from China in the form of a 'recipe', but was used to make objects of local style.

A further example of mapping alloy compositions, but on a finer scale, and using a different methodology, is shown in **Figure 7**. This shows the average value of tin in the metal assemblages for a range of sites across the European Circum-Alpine region during the first part of the Early Bronze Age (Perucchetti 2017). The interesting feature is a line running northwest to southeast across the Alps towards the western end, which we have termed the 'tin line'. To the west of this line, most of the sites have a high average value of tin, whereas to the east the averages are lower—generally below 1.5%. This suggests that the idea of tin bronze entered the Alpine region from the west rather than the east, and at this time did not cross the 'tin line'. Equally interesting, it also shows that the Alps themselves did not create a north-south barrier—tin is present in the north and south of the 'bronze' region, and is equally low north and south in the eastern region. This line must therefore represent a 'cultural barrier' to east-west movement.

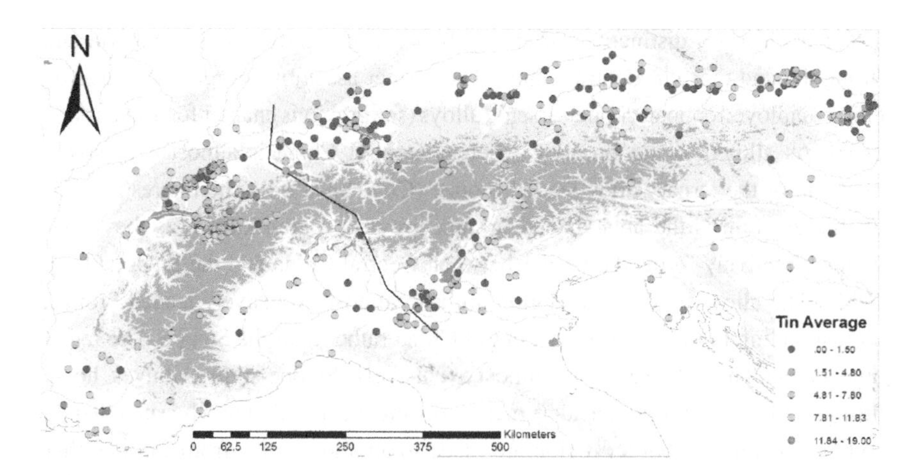

Figure 7:
Plot of the mean value of the tin concentration in objects from various sites in the Circum-Alpine region of Europe at the beginning of the Early Bronze Age (c. 2200–2000 BCE). The average values are coloured on a scale of <1.5% Sn to 12–19% Sn (Perrucchetti 2017, fig. 95).

A major advantage of our methodology with regard to both trace elements and alloying is that it allows us to consider the circulation of copper independently from what it is alloyed with. There has been a general tendency in previous studies to assume that alloys are immutable once formed (i.e., a tin bronze will always be a tin bronze). We make no such assumptions. We can imagine copper being alloyed and re-alloyed any number of times, perhaps in 'consumer' regions rather than in primary metal-producing areas. In our methodology we can continue to discuss the underlying copper chemistry (via copper groups) regardless of whether it is in the form of pure copper, or a leaded bronze, etc. This allows us to compare flows of different alloys in terms of their underlying copper chemistry (i.e., to think about whether different alloys are coming from the same copper source), and also to distinguish between the circulation of pre-alloyed metals and the practice of alloying (or re-alloying) in the consumer society. In effect, it is a powerful new tool which allows us to ask fundamental questions about the concept of alloys, and the roles and identity of alloyed metal in society.

To illustrate this, **Figure 8** shows some of the results of our re-interpretation of a small set of Qing dynasty coins (n = 38), taken from a study of Chinese copper alloy coinage, using in this case the chemical analyses by Wang *et al.* (2005, Appendix 1, p. 89). These data represent copper coinage of the early to middle Qing dynasty (*c.* 1644–1796 CE), and are chosen because they are well dated within this period.

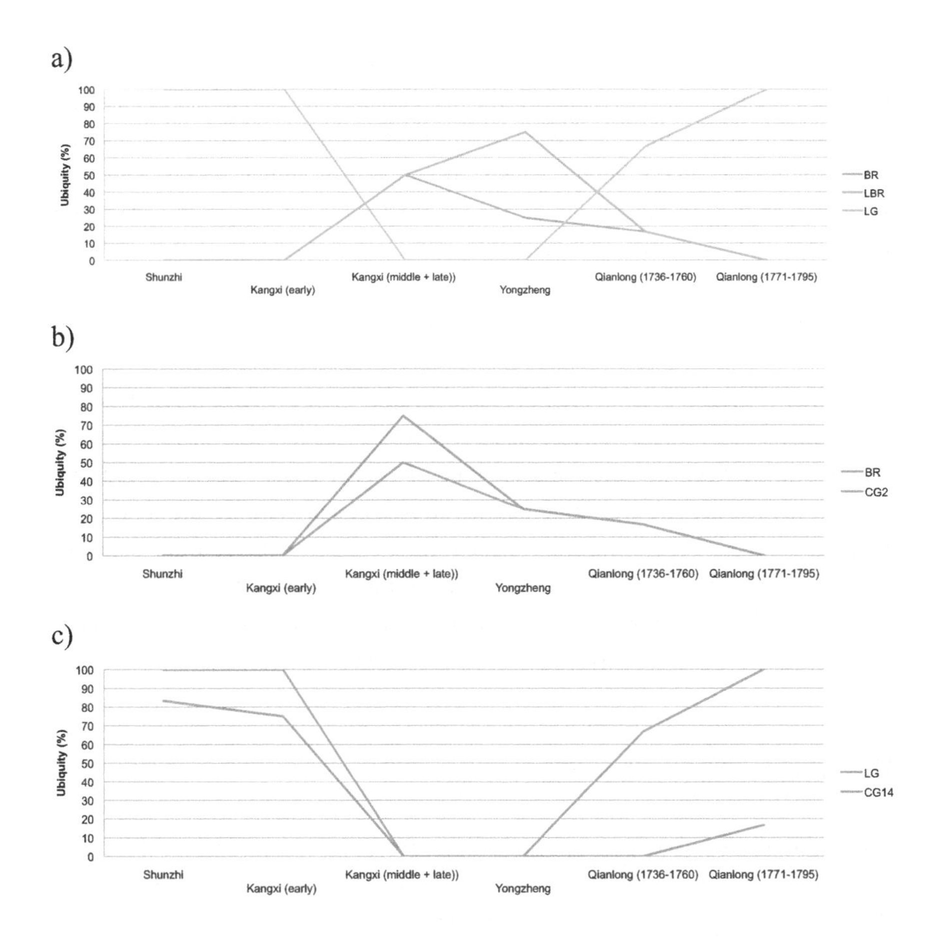

Figure 8:
Correspondence between copper groups and alloy types in Qing dynasty Chinese copper alloy coinage. 8a) The ubiquity of coins classified as Brass (BR), Leaded Brass (LBR) and Leaded Gunmetal (LG); 8b) CG2 and BR (Brass, Cu + Zn (>1%)); 8c) CG14 and LG (Leaded Gunmetal, Cu+ Sn (>1%) +Zn (>1%) + Pb (>1%)).

Although the numbers are small, Fig. 8a shows a remarkable switch in alloy type, from LG (leaded gunmetal: Cu + Sn + Pb + Zn, all >1%) during Shunzhi (1646–1653 CE) and early Kangxi (1682–1683/4) to BR (Brass, Cu + Zn (>1%) and LBR (Cu+ Sn (>1%) +Zn (>1%) + Pb (>1%) during the middle and later Kangxi period (1684–1723) and Yongzheng (1723–1736). From the beginning of Qianlong's reign (1736) there is a return to LG. What is more interesting, however, is the combination of these changes in alloy type with the copper group information for the same coins. Figure 8b shows a strong coherence between BR (Brass: Cu +Zn) and CG2 (Cu + As), and Fig. 8c shows partial coherence between the alloy LG and CG 14 (Cu + As, Sb and Ni, all >0.1%)—partial because in the early period they track each other closely, but diverge in Qianlong's reign. This shows that the appearance of brass coinage is closely associated with the use of copper containing only arsenic as the major impurity, which becomes the dominant coinage alloy in the later Kangxi and Yongzheng period (c. 1684–1736 CE). It is preceded, and also replaced, by a leaded gunmetal alloy which in the early period is strongly associated with copper classified as CG14 (Cu + As, Sb, Ni), but in Qianlong's reign, although the use of leaded bronze returns, it is no longer associated with CG14.

Further work and much more trace element data are required to understand the full implications of these correlations, but it would suggest that there is a close relationship between the alloy and the type of copper used. This could indicate that the alloy types used in Qing copper alloy coinage came to the mint from different sources, each using a different sort of copper. Although our database of Chinese copper coin compositions from several sources now contains several thousand analyses, dating from the Eastern Zhou to the late Qing, unfortunately only a minority of these analyses include trace element data.

The emergence of alloying

The origins of alloyed copper, especially tin bronze, have been studied and discussed by scholars for more than a century (e.g., Sullivan 1873, Smith 1935, Partington 1936), and there has been much consequent speculation about the sources of tin for the Bronze Age Mediterranean and Near East (e.g., Dayton 1971, Crawford 1974, Olin and Wertime 1978, Cleziou and Berthoud 1982, Muhly 1985). The aim of the discussion here is not to reconsider the evidence for identifying the time or place of the origins of alloying, nor particularly about the source of tin, but to think more about the spread of alloying as a practice across Eurasia.

The term 'alloy', and, more specifically, 'deliberate alloy', carries many implications which are somewhat complicated to interpret in an archaeological context, because they relate to understanding the *intentionality* of the metalworkers. The term alloy is of course adopted from modern materials science, and is taken to mean a mixture of metals which is designed to improve the physical or visual properties of the material. An important observation, apparent to anyone who has studied the chemical composition of Bronze Age metalwork, is that not all ancient compositions correspond to modern alloy definitions. Hence, as discussed above, we have adopted a more flexible approach to the definition of alloys, rather than simply applying modern definitions, and thereby implicitly assuming that ancient metalworkers were somehow groping towards modern alloy compositions. A key question in terms of determining intentionality is to learn at what point the alloying process happened within the metal production process. Knowing this would help us not only to understand the degree of control possible in producing alloys, but would also give us some insights into how alloyed metals were understood in the Bronze Age. In literate societies, considerable information on this later point can be obtained by studying the names used for the different metals and alloys (e.g., Bertholet 1888, Lepsius 1877), but for most of Eurasia this is not possible.

In modern metallurgical practice, alloying can easily be achieved by melting together weighed quantities of the alloying metals (allowing for predictable losses) so as to produce the targeted composition. This could certainly also have occurred in antiquity, especially with tin and lead, which potentially have been available as separate metals, and needed to be added in easily measured quantities of the order of 10–20% by weight. However, particularly in the case of alloys of arsenic and antimony, it may not have happened like this, because metallic arsenic was unknown, and metallic antimony was rare. Even if these metals were available, the quantities needed to create an alloy (up to a few percent) might suggest that direct mixing was unlikely. The addition of zinc to produce brass was definitely more complicated, because, as with arsenic, metallic zinc was largely unknown in antiquity and is also volatile, requiring special conditions to produce the alloy (Pollard *et al.* 2017b, 250–255).

It is possible that the desired alloys were obtained directly by smelting specific polymetallic sulfide ores, defined in modern parlance as a mineral deposit with three or more metals in commercial quantities, commonly including Cu, Pb, Zn, Fe, Mo, Au, and Ag. Plausibly, as has been argued by Radivojević *et al.* (2013) for the Balkans, some form of co-smelting of such ores may have been the way in which alloying copper with tin was first discovered. An alternative route for certain regions may have been the smelting of secondary deposits from oxidized tin sources, which can produce copper-rich minerals such as mushistonite (Cu, Zn, Fe^{2+})[$Sn(OH)_6$], the type site of which is Mushiston, in Tajikistan. Such secondary minerals are green, and it possible that in the early stages of copper smelting they were mistaken for pure copper minerals (such as malachite, $Cu_2CO_3(OH)_2$), but, when smelted, would have produced a natural tin bronze.

Apart from the single smelt of such a natural mixed ore, it is also possible to produce alloys by co-smelting ordinary copper ores with selected ores containing the required elements. Equally, the re-smelting of already smelted copper with ores containing the alloying metal could produce an alloy. In such cases, there is clearly an intention to produce a metal with desired properties, but they are not best described as 'alloying' in the modern sense of the word. They are practices that are more related to the experience of the smelter, who would not have known the chemistry of the ores, but would have known that ores from certain places, or particular mixtures of ores from specific places, produced metals with particular characteristics. This is another case where the unquestioning application of modern terminology can hide the true nature of the process, and also the intentions of the ancient metalworkers. A final route to an alloy would be the mixing of metals, some of which might contain a percentage of alloying metals. In such cases, the alloying metal(s) would be diluted, and might have little effect on properties or appearance. This could indicate recycling, as well as the absence of intention to produce an alloy, and is perhaps the reason why many archaeological alloys have intermediate compositions.

Such considerations are especially important when considering alloys of copper with arsenic and antimony, but this is complicated by the volatility of arsenic (and to a lesser extent of antimony). As with tin and lead, copper intentionally alloyed with arsenic or antimony (probably by co-smelting copper- and arsenic-ores, or by smelting polymetallic sulfides) should show a dominant peak in the distribution at a measurably high value—probably greater than 3–5%—but with a tail towards lower values caused by the largely uncontrollable losses due to volatility. A distribution dominated by lower values close to zero but with few higher values is more likely to be the result of accidentally smelting ores containing arsenic or antimony rather than a deliberate intention to produce arsenical or antimonial copper. We must also note that, as shown in the previous chapter, arsenic and, to a lesser extent, antimony, are volatile if copper is heated in oxidizing conditions, meaning that any subsequent re-melting may lower the concentrations of arsenic and antimony.

We can use the FLAME database to look for 'hotspots' across Eurasia in the use of alloys containing significant amounts of arsenic. **Figure 9** shows that there are a limited number of regions (plotted as 1^0 WGS84 grids) where the ubiquity of objects containing >1% arsenic is greater than 50%. These include Iberia, the Caucasus into central Asia and the region around the Altai Mountains, which would highlight these areas as being those where Cu-As alloys may have been deliberately produced. It must be remembered, however, that the database is not yet complete, and that this figure shows all periods.

Similarly, by mapping the ubiquity of tin bronzes within regional assemblages, it is possible to chart the development and spread of the use of tin bronze, and even to indicate where the tin itself might be coming from. **Figure 10** shows tin 'hot-spots' focussed on western and central Europe, Afghanistan and the Caucasus, and China. The first three are well-known locations for tin mines, and China must also have had a major source of tin. A similar figure for lead (**Figure 11**) shows, however, that China is almost unique for focussing on leaded bronze as the primary alloy.

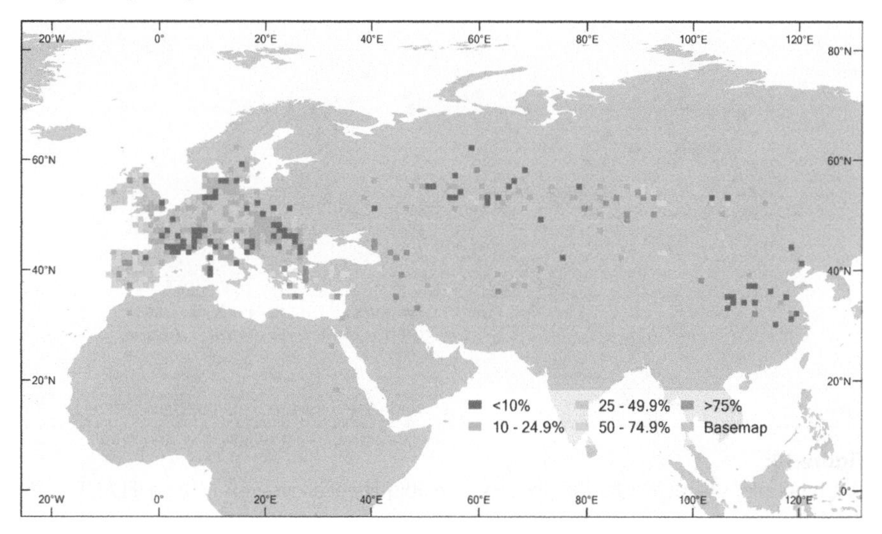

Figure 9:
Distribution of sites across Eurasia showing the ubiquity of alloys with >1% arsenic (FLAME database).

It is possible to use the same data as shown in Figure 10 to look for the *early* locations of the widespread adoption of tin bronze and its subsequent spread, by plotting the ubiquity of tin bronze on a series of time resolved maps (**Figure 12**). (We cannot look for the *earliest*, because these occur in the Late Neolithic or Eneolithic levels in the Balkans and perhaps elsewhere, and as such have not been captured within the FLAME database). From the data we have assembled so far, this shows the earliest focus (c. 3500 BCE) to be in Anatolia and the Danube basin, with some examples in central France. Between c. 3500 and 3000 BCE this spreads to Germany and some other parts of western Europe, but by c. 3000–2500 BCE the focus shifts dramatically to the Atlantic coast of Europe, potentially signalling the beginnings of the exploitation of Cornish tin. Some occurences are also seen around the Urals and northern Kazakhstan. The picture is unchanged until c. 2500–2000 BCE, when the southern Urals emerge as a strong focus for tin bronze. By c. 1500–1000 BCE, this focus moves eastwards towards the Altai and southern Siberia, and also China emerges as a major centre. This picture is certain to evolve as more data and better chronologies are included, but it illustrates the power of mapping the ubiquities of the various alloys.

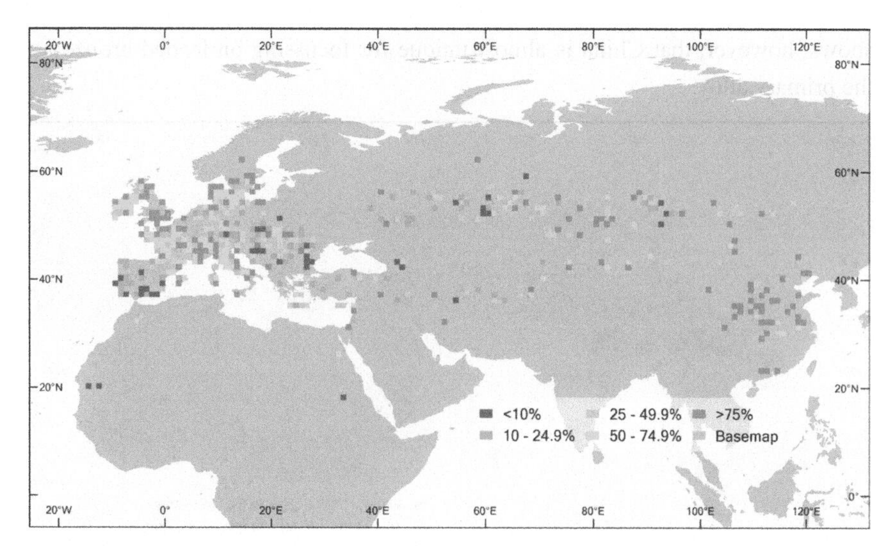

Figure 10:
Distribution of sites across Eurasia showing the ubiquity of alloys with >1% tin (FLAME database).

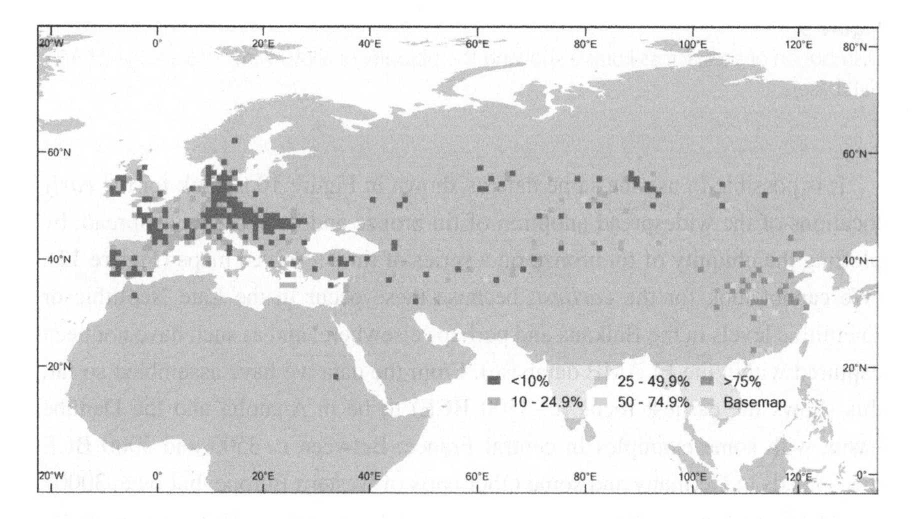

Figure 11:
Distribution of sites across Eurasia showing the ubiquity of alloys with >1% lead (FLAME database).

It is worth noting that, even if, as may be the case, these 'hot-spots' do not exactly coincide directly with known tin sources, they do give us information about where tin bronzes were actually being used, from which we can infer the social context in which this was occurring, and which objects were preferentially being made of bronze. This may be at least as important as knowing where the tin itself comes from.

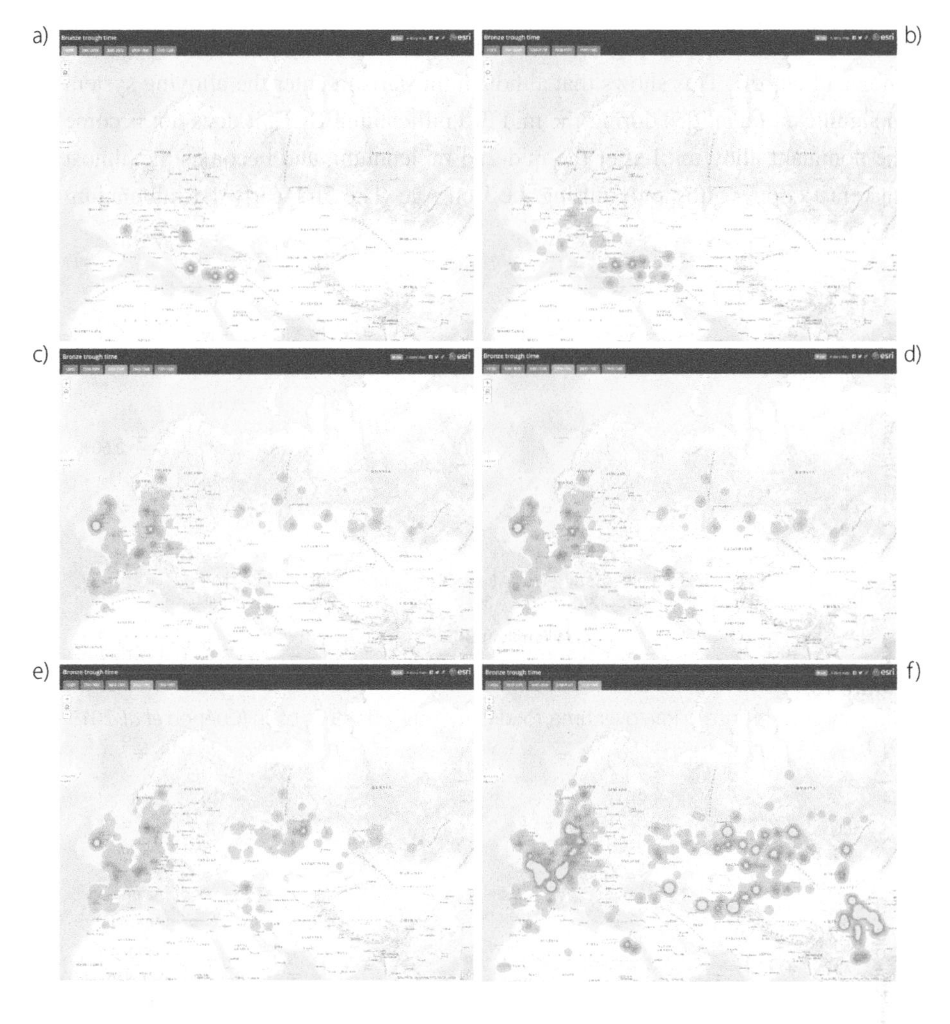

Figure 12:
A series of time slices for the ubiquity of tin bronze (Sn >1%) across Europe. a) c. 3500 BCE; b) c. 3500–3000 BCE; c) c. 3000–2500 BCE; d) c. 2500–2000 BCE; e) c. 2000–1500 BCE; f) c. 1500–1000 BCE (FLAME database).

It is obviously also important to look at the adoption of tin bronze on a finer scale in specific regions, and also not to just look at the timing of the spread of the alloy, but also at the uses to which it was first put. This was done for Iran and adjacent regions by Cuénod (2013), and is being done for a wider area of Western Asia by Howarth (forthcoming). **Figure 13** (Cuénod *et al.* 2015) shows the ubiquity of tin bronzes in all Iranian objects as a function of date, from the 5th to the mid-1st millennium BCE. We have used our definition of tin bronze (i.e., Cu + Sn >1%, Pb and Zn <1%), but in this figure we have also varied the cut-off for tin, using

0.5%, 1% and 2%, showing that the overall pattern is largely insensitive to the choice of cut-off. This shows that although tin starts to enter the alloying system in significant quantities during the mid-3rd millennium BCE, it does not become the dominant alloy until after the mid-2nd millennium, and becomes the almost universal copper alloy only during the Iron Age (late 2nd–early 1st millennium BCE).

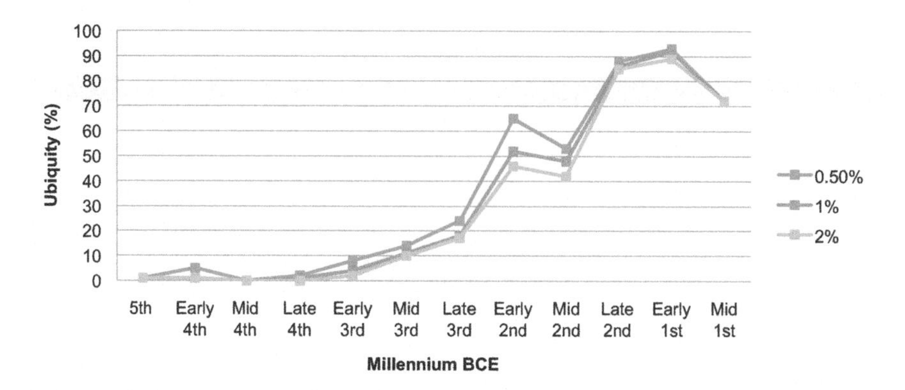

Figure 13:
Ubiquity of tin bronze in Iran over time, for different cut-off values of Sn (Cuénod ***et al.*** 2015, fig 4).

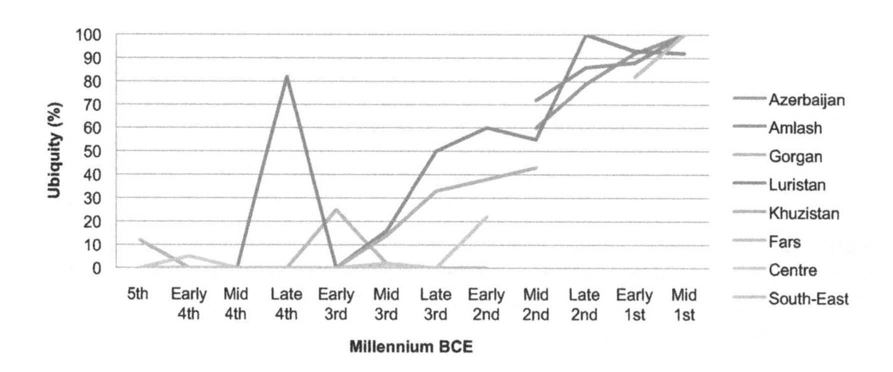

Fig 14:
Ubiquity of tin bronze in different regions within Iran over time (Cuénod 2013, fig. IV-3, p. 161). Sn cut-off set at 2%.

In her thesis, Cuénod (2013) explored the spread of tin bronze into Iran on a regional basis (**Figure 14**). This shows that most regions within Iran follow the general pattern outlined above, but in one region (Luristan) the arrival of tin bronze, although transient, was considerably earlier than elsewhere. The role of

Luristan (and possibly Deh Hossein) in the supply of tin was discussed above (Figure 5), but the clear implication is that Luristan exploited a local source of tin earlier than the rest of Iran.

In terms of the first use to which tin bronze was put, Cuénod (2013) divided up the objects from Iran into broad functional categories (weapons, tools, ornaments and vessels, the latter not plotted because there are too few). Plotting the ubiquity of tin bronzes over time by category showed little variation in the adoption of tin bronze across these categories (**Figure 15a**). In contrast, the same analysis for Mesopotamian objects strongly suggests that tin bronze was first used for vessels (**Figure 15b**). Because the use of tin bronze for tools and weapons came later, this would suggest that the property most desired in tin bronze in Mesopotamia was the colour for display, rather than any considerations of the physical properties of the alloy.

a)

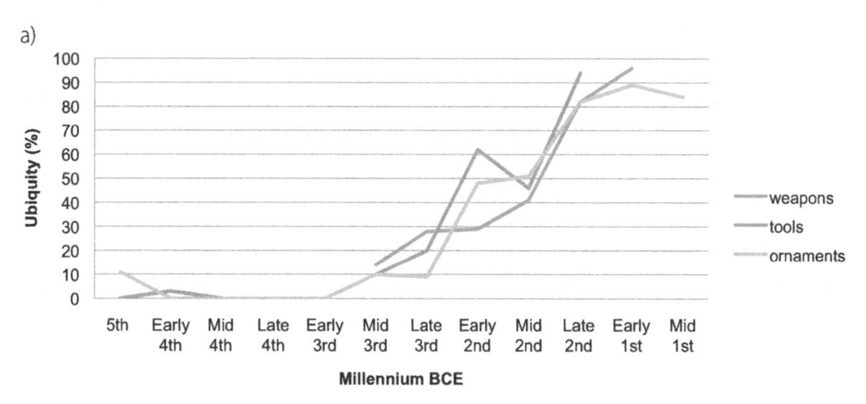

b)

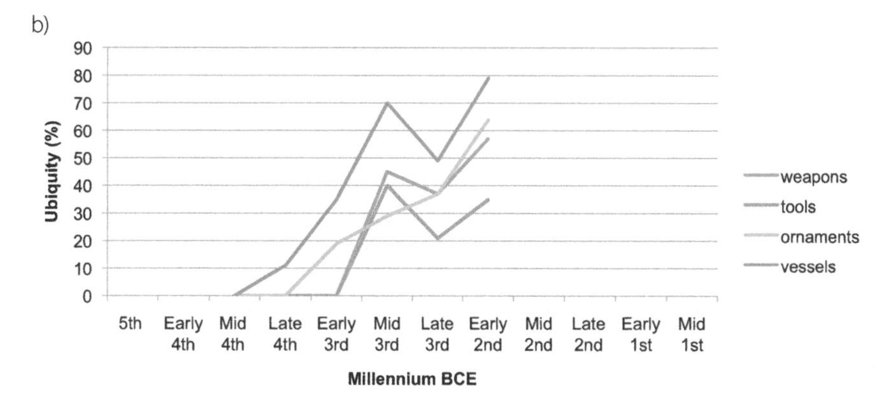

Fig 15:
Ubiquity of tin bronze in a) Iran and b) Mesopotamia over time, broken down into broad functional categories (Cuénod *et al.* 2015, fig 6).

Alloy profiles, cumulative frequency distributions and 'regional alloying practices'

Figure 3 (above) shows a simple example of comparing the distribution of tin in two Chinese Shang dynasty assemblages—that of Fu Hao's tomb, and the tombs in the Western Area of Anyang (Liu 2016), showing a distinct difference in the profiles. Normally the distributions of alloying elements in different assemblages are compared simply by using histograms of the raw data. These can take the form of a sequential ranked set of bar charts of the alloy content in each object, or more commonly as histograms of the number of objects whose alloying element concentration falls within a particular range of compositions (the 'bin width'). In order to make the comparison between assemblages of different sizes easier, we usually convert the number of objects in each 'bin' into a percentage of the total number of objects in the assemblage. Normally, as in Figure 3, we tend to present these as distribution profiles by drawing a line joining the tops of each bar in the histogram, usually displaying only the line and not the bars (a *skyline* plot). An example of the lead concentration in a range of Chinese, Mongolian and Eastern Siberian assemblages is shown in **Figure 16**, in which, for example, the number of objects from Minusinsk containing <1% Pb is approximately 97% of the total number in the Minusinsk assemblage. Almost none of the Minusinsk objects have more than 7% Pb, whereas only 27% of the Central Chinese objects contain less than 1% Pb, and approximately 27% have more than 10% Pb.

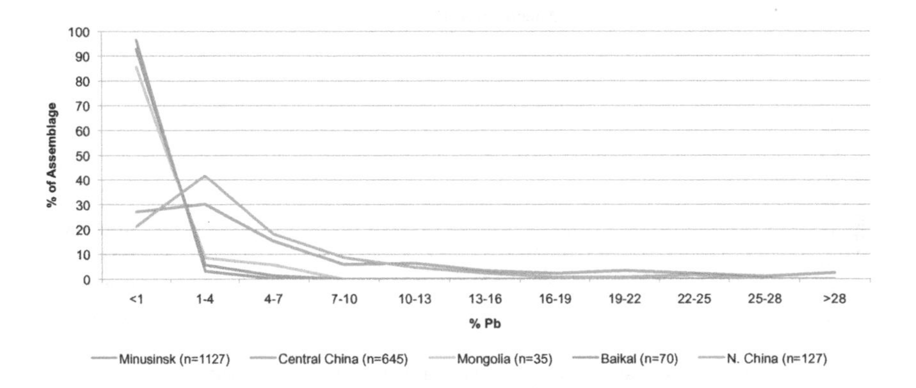

Figure 16:
Comparison of lead distribution in a range of East Asian metal assemblages dated to the second half of the second millennium BCE (after Hsu 2016).

It is clear from this figure that there are different patterns of lead addition according to region, at least as represented by the available assemblages. In this example, we might reasonably conclude that there are similarities between the practices of lead addition in Mongolia, Baikal and the Minusinsk basin, and that these differ from those used in Central China and North China (Hsu 2016, 228–230).

However, it is difficult to go beyond a simple visual comparison using these curves because:

i) histogram presentation is always subject to arbitrariness in the choice of bin width and interval;

ii) differentiating between two profiles is essentially subjective. In particular, there is no allowance for the influence of different sample sizes when comparing curves.

The 'bin-width' issue in histogram presentation is well-known. There is no infallible rule for choosing the number and width of bins, but there are several ways of calculating the most appropriate number for particular datasets, depending on the number of samples in the distribution. The simplest 'rule of thumb' is to say that the number of bins should be the square root of the number of data points. Thus an assemblage containing 100 objects should be plotted on a histogram with 10 equal bins, which must cover the entire range of values. In Figure 16 above there is another issue, in that each region is represented by a vastly different number of objects (from 35 for Mongolia up to 645 for Central China), and hence the optimal bin width will differ between regions. In such a case, the only practical course of action is to calculate the bin width from the dataset with the lowest number of objects and use this for all the distributions.

The visual effect of the choice of bin width is illustrated in **Figure 17**. The data presented are a simulated dataset of 100 objects containing tin between 0 and 3%. Figure 17a shows the allocation of objects to bin, for two different bin widths of 0.1% and 0.3%. As would be expected, the distribution at 0.1% bin width shows more detail in terms of fluctuations and gaps in the data, whereas the 0.3% bin width shows a smoother and more continuous distribution. According to the rule of thumb described above, the optimal number of bins for these data is 10, so the 0.3% bin width should be regarded as the most useful presentation. Figure 17b shows the effect of presenting these same histograms in the form of a skyline plot, in which the tops of each column are simply joined up. It is clear from these two figures that the choice of bin width can have a significant effect on the visual appearance of the histogram, and hence on the impression we take away from the distribution of the data. In Figure 17b, for example, the 0.3% bin presentation gives a strong impression of a preponderance of the data falling into the lowest bin (0–0.3%), plus two 'peaks' around 0.9–1.2% and 1.5–1.8%, which are less easy to appreciate in the 0.1% bin data.

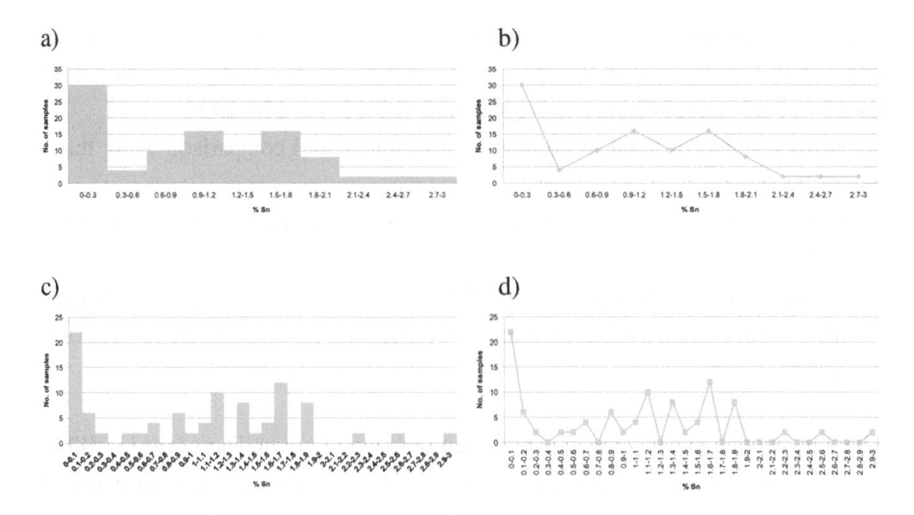

Figure 17:
Histogram of same dataset with a) bin width of 0.3% and c) bin width 0.1%. Skyline plots of same data at b) 0.3% and d) 0.1% bin widths.

Cumulative Frequency Distribution Functions and the Kolmogorov-Smirnov Test

An alternative way of presenting the distributional data of alloying additions to copper is to calculate the cumulative frequency distribution functions (CDF). This is calculated from the elemental histogram profiles, such as those shown in Figure 17, simply by starting from the left hand side and replacing the number of counts in each individual column with the sum of counts in that column plus all previous columns, and expressing the total as a percentage of the total number of samples. The CDF curve is then plotted on the same axis as the original histogram, showing a profile of how the data accumulate from zero up to the maximum percentage represented. An example is given in **Figure 18**, where the left hand vertical axis relates to the histogram (blue) presentation and is the number of objects per bin, and the right is the proportion of the total assemblage, relating to the CDF (red). Following on from the discussion of the effect of different bin-widths on the appearance of histograms in the previous section, **Figure 19** shows the CDFs for the simulated data allocated to two different bin widths presented in Fig 17a. The CDFs for the same data but at two different bin widths are virtually identical, showing that this form of presentation is much less susceptible to distortion by the selection of bin width than is the conventional histogram.

This form of presentation also has the major advantage that it allows two distributions to be statistically compared using the Kolmogorov-Smirnov (KS) test, which assesses the probability that two distributions could have been drawn at random from a single parent distribution. The observed difference is unlikely to have occurred by chance if the KS test statistic exceeds a calculated critical value, and under these circumstances the two distributions can therefore be considered to be significantly different from each other.

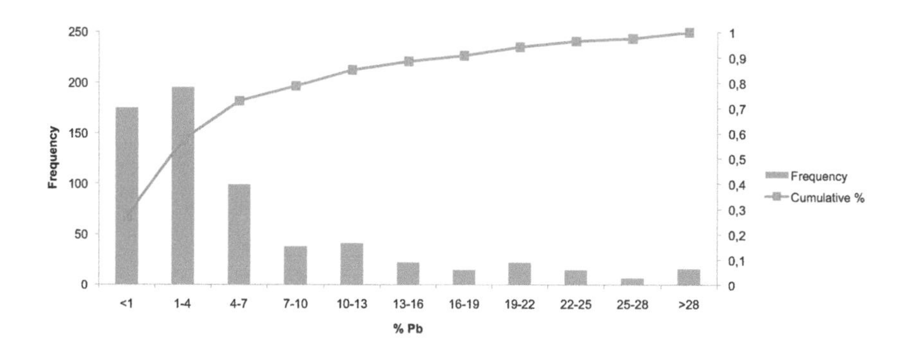

Figure 18:
An example of the cumulative frequency distribution function: histogram and cumulative frequency distribution function for the Pb content of Central Chinese bronzes as shown in Figure 16 (FLAME database, after Hsu 2016).

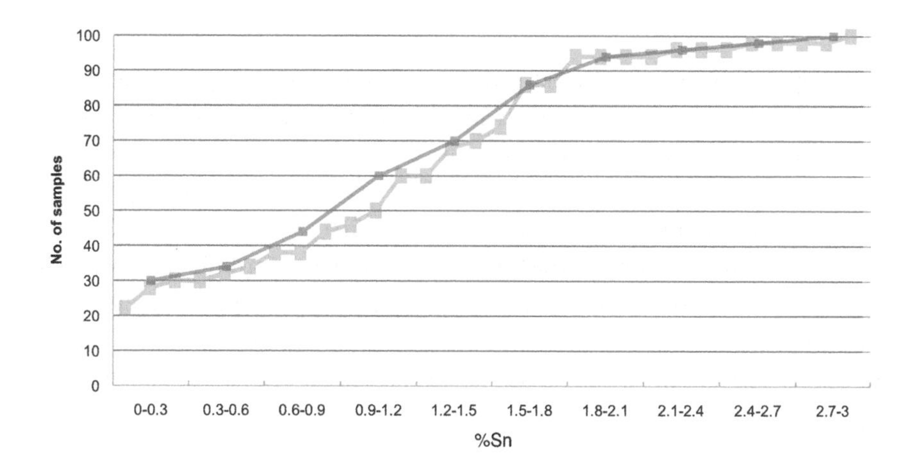

Figure 19:
Cumulative frequency distribution function of simulated data in fig. 17. The blue line is for a bin width of 0.1% and green for 0.3%.

In order to demonstrate this methodology and the application of the KS test, we give a worked example. We (and many others) have studied the patterns of continuity and change between the composition of Chinese Shang (c. 1600–1046 BCE) and Western Zhou (1046–776 BCE) bronzes – in particular in the practice of adding lead (e.g., Pollard *et al.* 2017a). **Figure 20** shows the cumulative frequency distributions of lead in bronzes during the Anyang period of the Shang dynasty (c. 1200–1046 BCE), separated into the four chronological phases identified at Anyang, namely Yinxu I to IV (Jin *et al.* 2017).

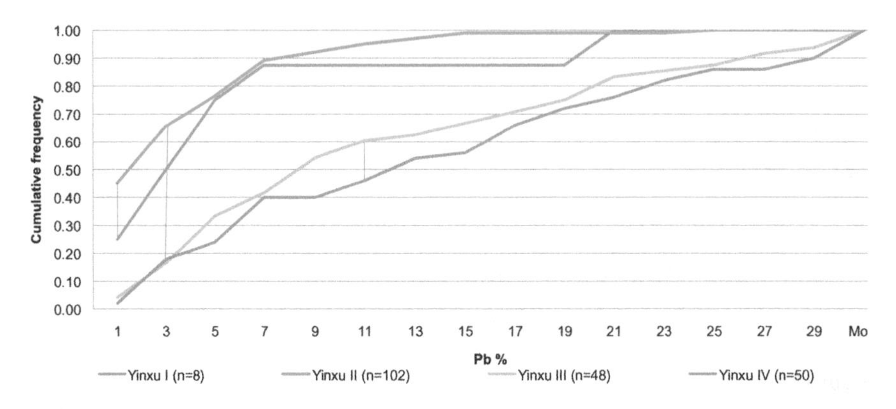

Figure 20:
Cumulative frequency distribution for lead in the Shang Anyang period, divided into Yinxu phases I–IV (Data from Zhao 2004). Vertical lines between curves show points of maximum difference between YI and YII, YI and YIII, and YIII and YIV.

There appear to be two different patterns within these data, namely that Yinxu phases I and II are similar, as are Yinxu III and IV, but these two patterns are different from each other. The Kolmogorov-Smirnov test allows us to test whether these apparent differences are more than might be expected by chance, given the different number of samples in each period. The first step is to tabulate the data shown in Figure 20 (**Table 2**), which then allows the maximum differences between any two distributions to be simply calculated. The differences are measured vertically between equivalent points on each curve, and represent the maximum difference between two distributions for any fixed value of lead concentration.

Bin	Yinxu I	Yinxu II	Yinxu III	Yinxu IV	D (YI-Y2)	D (YIII-YIV	D (YII-YIII)
1	0.250	0.451	0.042	0.020	-0.201	0.022	0.409
3	0.500	0.657	0.167	0.180	-0.157	-0.013	0.490
5	0.750	0.765	0.333	0.240	-0.015	0.093	0.431
7	0.875	0.892	0.417	0.400	-0.017	0.017	0.475
9	0.875	0.922	0.542	0.400	-0.047	0.142	0.380
11	0.875	0.951	0.604	0.460	-0.076	0.144	0.347
13	0.875	0.971	0.625	0.540	-0.096	0.085	0.346
15	0.875	0.990	0.667	0.560	-0.115	0.107	0.324
17	0.875	0.990	0.708	0.660	-0.115	0.048	0.282
19	0.875	0.990	0.750	0.720	-0.115	0.030	0.240
21	1.000	0.990	0.833	0.760	0.010	0.073	0.157
23	1.000	0.990	0.854	0.820	0.010	0.034	0.136
25	1.000	1.000	0.875	0.860	0.000	0.015	0.125
27	1.000	1.000	0.917	0.860	0.000	0.057	0.083
29	1.000	1.000	0.938	0.900	0.000	0.038	0.063
More	1.000	1.000	1.000	1.000	0.000	0.000	0.000
n=	8.000	102.000	48.000	50.000			
				D	0.201	0.144	0.490
				Dcrit	0.367	0.202	0.175
					Same	Same	Different

Table 2:
Tabulated cumulative frequency data for lead from Figure 20.

The KS test statistic is calculated from the maximum absolute difference (D) between the two distributions for all values of x (absolute difference means that we ignore the positive or negative nature of the difference, and just take the number). From Table 2, the maximum absolute difference D between the distributions for Yinxu I and II is 0.201 (column labelled 'D (YI–YII)', maximum value shown in red). In order to test whether this difference D could have been obtained by chance, it is tested against a critical value (D_{crit}), which depends on the number of samples in each distribution (n_1 and n_2), and is given by:

$$D_{crit} = c(a)\sqrt{\frac{(n_1 + n_2)}{n_1 n_2}}$$

where $c(a)$ depends on the level of confidence required (normally 95%, in which case $c(a)$ is 1.36). In this example, $n_1 = 8$ and $n_2 = 102$, and so D_{crit} is 0.367. Since the observed maximum difference between Yinxu I and II (0.201) is less than the critical value (0.367), it is likely that the distributions for Yinxu I and II differ just by chance, and the difference is therefore not significant. We can thus conclude that, as might be expected, the Pb profiles in bronzes from these two periods are the same, and therefore that the lead alloying practices of Yinxu I and II periods are indistinguishable, based on the available data. In fact, in this example, the value of D_{crit} is unusually high, because one of the distributions contains only eight samples. Normally we would expect to use much larger numbers of objects in each distribution—ideally, with $n_1 \approx n_2 > 50$.

Following the same procedure, Yinxu III and IV can be shown to be indistinguishable ($D = 0.144$, $D_{crit} = 0.202$), but Yinxu II and III are demonstrably different, since $D > D_{crit}$ ($D = 0.491$, $D_{crit} = 0.175$). Thus there is a change in lead alloying practice between Yinxu phases I and II and Yinxu phases III and IV, which corresponds to higher levels of lead being used in Yinxu III and IV. Although this difference has been noted before from the raw histograms by other authors (e.g., Li 1982; Li *et al.* 1984; Liu *et al.* 2007; Zhao 2004; Zhao *et al.* 2008), we suggest that the cumulative frequency diagrams *show* these differences more clearly, and, crucially, allow the significance of such differences to be statistically tested.

It is obvious, however, that the differences we have shown between Yinxu I and II and Yinxu III and IV strictly only apply to the samples represented in the assemblage, not to all the objects produced or recovered during these periods. If the samples selected are not representative of the parent population, then the conclusions drawn are likely to be suspect. This is not, of course, a problem unique to the process described here—it is applicable to any work carried out on archaeological metals, and, indeed, more generally in all of archaeology. The samples used here are, at best, a biased sample of a biased sample of a biased sample from an unknown parent population! The unknown parent population consists of all the metal objects produced at Anyang during, for example, the Yinxu I period. The three levels of bias referred to depend on:

i) the objects which entered the archaeological record *during* Yinxu I, which is likely to be a small and unrepresentative sample of what was available;
ii) the objects *recovered* archaeologically from Anyang dating to Yinxu I, which depends on factors such as where the archaeologists choose to excavate, and
iii) the objects selected for chemical analysis from those excavated, which depends on availability and the interests of the analysts in terms of object typology.

If we knew what the parent population 'demography' was (i.e., the total number and type of all objects produced), then we could apply a correction to some of these biases, as is done with modern opinion poll data, but of course we do not. The best we can do is to be aware of their existence, and take what steps we can to minimise them, largely by ensuring that the samples included in the assemblages are as well-controlled as possible typologically and contextually. In practice, what this means is that when carrying out comparisons of the type illustrated above, we must explicitly justify the selection of samples in each assemblage in terms of how they directly address the specific archaeological question being asked.

Regional alloying practice

Keeping in mind the above concerns, we can, however, take the use of CDF's one stage further by suggesting that they might be regarded, under certain circumstances, as a reflection of the 'alloying practice' or 'alloying tradition' in a region (Pollard *et al.* 2018b). By this we mean the range of copper alloy compositions typically produced by the casting foundries within a defined region and period. This is similar to the idea of 'metallurgical focus' introduced by Chernykh, and defined by him as "*a region where similar metals and metal artifacts were produced professionally by a distinct group of skilled craftsmen*" (Chernykh 1992, 7). We can perhaps think of the regional alloying practice as representing 'casting foundry recipes' on a regional scale, in which case they become a way of linking individual alloy compositions to foundry practice (Kuijpers 2018, 242–253). We would of course expect a wide range of alloy compositions to be produced within a particular region, according to quality and typology—high-quality vessels with one target composition, mirrors another, weapons yet another, and so on—but we might also expect certain commonalities of practice, perhaps as a result of centralized control, or just from shared knowledge and 'custom and practice'. If we believe that the assemblage we have constructed from the available data is representative of the entire metal production of a region, then we can justifiably refer to it as a *regional alloying pattern*. This might be a reasonable assumption if we have a very large number of samples from across the spectrum of typologies (or, conversely, might be the only practical course of action, if we have assembled all the analyses we can find), but normally it is unlikely to be completely realistic. It is more likely to be meaningful if we restrict the scope of the assemblage in some way. For example, we might narrow the definition of the parent population away from 'all the objects produced in a region' to 'all the knives of type x produced in a region', or 'all the high status objects produced in a region'. In such cases we cannot then consider the assemblage to represent a '*regional* alloying pattern', but nevertheless by careful consideration such concepts and comparisons can be made archaeologically meaningful and useful.

We can consider comparisons of regional alloying practice across either space or time, or both, depending on how the assemblages to be compared are defined. Each cumulative frequency curve reflects the composition of the total range of alloys produced in one specific region or period, at least in so far as is evidenced by the specific assemblage used to construct that curve. When two regional or temporal distributions can be shown to be statistically indistinguishable, then the alloying element profiles can be said to share a common pattern. We would like to interpret this as implying that the two regions shared a common alloying practice. For this to be strictly meaningful it requires careful matching of typology, object size, and chronology between the two assemblages, as discussed above; however, if the data sets are large enough this interpretation may still be useful. Thus, in Figure 16, we can deduce (and could demonstrate statistically) that there was some similarity between the lead distribution in the assemblages from Central and North China on the one hand, and Transbaikal, Cis-Baikal and Minusinsk on the other, with Mongolia somewhere in between. We would then argue that this pattern represents some form of technological connectivity within these groups of regions—the 'Chinese' and the 'Steppe' traditions of metal-making respectively—but with technological boundaries between them. This then becomes a useful tool for comparing and contrasting the perceptions and use of metal between different cultures.

During the course of our work with these ideas, we have come across a situation where, say, the tin profiles are indistinguishable between regions but the lead profiles in the same assemblages are not. This could be telling us something about what is controlled and what is not controlled in the alloying processes in the two regions—in this example, is the tin controlled, but the lead added more randomly, but at the expense of the copper, not the tin? This might imply that there is a commonality in the use of tin in the two regions, but that lead is treated differently. If so, this could provide interesting insights into differing foundry practices, or different organization of the metal supply networks. When considering such situations, however, we need to remember that for an alloy consisting primarily of lead and tin added to copper, we have a version of the well-known *unit sum problem* (Aitchison 1986), i.e., that Cu + Sn + Pb ~ 100%, so that adding, say, more lead automatically reduces the concentration of both copper and tin, which will hence affect the CDF of tin in the assemblage. This requires more work and thought, and it could also be that using the absolute weights of each metal in the object would be more informative than the relative percentages. This would of course require the weights of all the objects to be known.

CDFs therefore appear to offer a powerful tool for comparing elemental distribution profiles from assemblages of metal objects from different regions or time periods—both in terms of the clarity of presentation, and also in the ability to statistically test whether such distributions are the same or different. A deeper question is what is the significance of these 'regional' patterns of alloying, and is the idea of 'regional alloying patterns' meaningful?

Kernel Density Estimates (KDE)

An alternative approach to presenting the distribution of alloying elements in an assemblage as histograms is to use kernel density estimates (KDE), which then enables them to be shown as quasi-continuous distributions. In fact, because Figure 16 above is a skyline plot, it appears to present the data as quasi-continuous distributions, but is in reality only a set of histograms. The use of KDEs does not actually remove the fundamental arbitrariness of the selection of bin width associated with histograms, since the width of the selected kernel is obviously also arbitrary, but equally there are formulae for estimating this based on the number of data points. In fact the KDE renditions are more than simply smoothed quasi-continuous representations of the data—they can be considered as probability distributions representing the underlying data. They then have the property that the area beneath the curve is normalised to 1 (or 100% probability), and so the similarity between two distributions can be estimated by calculating the degree of overlap between them. This gives an alternative quantitative approach to the CDF methodology described above for comparing distributions. We return to this use of KDEs when considering lead isotope data in the next chapter.

Chapter 6

Lead Isotope Data from Archaeological Copper Alloys

Isotopes are atoms which possess the same chemical characteristics (by having the same number of protons in their atomic nucleus) but have different atomic weights because of a variable number of neutrons in the nucleus. Most elements exist as more than one isotope: for example, carbon exists as ^{12}C (six protons plus six neutrons) and ^{13}C (six protons plus seven neutrons), with an average abundance ratio of 99:1. Apart from the light elements (those lighter than oxygen), which can be significantly fractionated by biological processes, most elements show relatively little variation in the abundances of their natural isotopes. For the heavier elements this variation arises largely from geological processes (such as different emplacement temperatures during mineralization). The exceptions tend to be elements which have one or more isotopes produced by the radioactive decay of other elements. Lead is one of the more unusual in that it has a very wide range of natural isotopic variation, due to the fact that three of its four stable isotopes (^{206}Pb, ^{207}Pb, and ^{208}Pb) lie at the end of major uranium and thorium radioactive decay chains. These chains with their respective half-lives are as follows:

$$^{238}U \rightarrow {}^{206}Pb \qquad T_{1/2} = 4.468 \times 10^9 \text{ years}$$
$$^{235}U \rightarrow {}^{207}Pb \qquad T_{1/2} = 0.7038 \times 10^9 \text{ years}$$
$$^{232}Th \rightarrow {}^{208}Pb \qquad T_{1/2} = 14.01 \times 10^9 \text{ years}$$

Thus the abundance of these three isotopes of lead present in a deposit can increase dramatically depending on the amount of U and Th present. The fourth stable isotope, ^{204}Pb, is not produced by radioactive decay but is residual from the formation of the universe, and is often therefore termed *primeval*. The accepted average natural abundances of the four stable isotopes of lead are ^{204}Pb = 1.4%, ^{206}Pb = 24.1%, ^{207}Pb = 22.1%, and ^{208}Pb = 52.4%, but the variation can be large. Russell and Farquhar (1960, 14) give examples of the ranges for each isotope abundance as follows: ^{204}Pb = 1.044 – 1.608%, ^{206}Pb = 21.53 – 28.39%, ^{207}Pb =

19.22 – 23.46%, and ^{208}Pb = 51.2 – 53.40%. They even quote an extreme example of lead from a uranium mine with ^{204}Pb = 1.044%, ^{206}Pb = 41.87%, ^{207}Pb = 19.45%, and ^{208}Pb = 37.64%.

Although this range of natural isotopic variation is, relatively speaking, extremely large, it is usual to record and report lead isotopes as a set of three isotope ratios, since this allows greater precision in the measurements. Geologically, they are usually reported as $^{206}Pb/^{204}Pb$, $^{207}Pb/^{204}Pb$ and $^{208}Pb/^{204}Pb$, but archaeologically the first results were published as $^{206}Pb/^{204}Pb$, $^{206}Pb/^{207}Pb$ and $^{208}Pb/^{207}Pb$ (Brill and Wampler 1967). Subsequent archaeological practice has tended to report $^{204}Pb/^{206}Pb$, $^{207}Pb/^{206}Pb$ and $^{208}Pb/^{206}Pb$, originally because ^{204}Pb is the least abundant (~1.4%) and therefore ratios measured against ^{206}Pb (~24.1% abundant) are capable of higher measurement precision. In the geochemical literature it is well known that modern magmatic and ore provinces tend to form very narrow alignments in plots of $^{207}Pb/^{206}Pb$ versus $^{208}Pb/^{206}Pb$, which overlap strongly (e.g., Albarède *et al.* 2012) and therefore have limited resolution between ore sources. For these reasons, we prefer to use the ratios as reported in the geological literature, i.e., as ratios to ^{204}Pb. It is a simple procedure to mathematically convert between one set of ratios and another, but the quantities so produced are not necessarily equivalent. There are clearly seven possible ratios of the four isotopes, which are completely *mathematically* interchangeable. However, the practicalities are that in measurement by both TIMS (thermal ionization mass spectrometry—the original method) and ICPMS (inductively-coupled plasma mass spectrometry) (see Pollard *et al.* 2017b, 81–82) the data are actually recorded directly as three specific ratios rather than four independent abundances. Switching from the ratios as measured to another set therefore increases the error term associated with the calculated ratio, in a way which can be difficult to predict. Technically, therefore, the ratios should not be re-calculated from those which were originally measured (or, to put it another way, it *does* matter which ratios are quoted). However, the increase in the error term is only likely to affect the fourth or fifth decimal place of the reported ratio, and, as we show below, we are often only considering isotope ratios to the first or second decimal place. So, in our work, the conversion from measured ratio to a ratio against ^{204}Pb, if necessary, should have very little effect.

Brill and Wampler (1967) showed that by using measurements of the lead isotope ratios, it was possible to differentiate lead ores (galena, PbS) coming from Laurion in Greece from those occurring in England and Spain. They did note that an ore sample from north-eastern Turkey fell into the same "isotope space" as that occupied by three ores from England, thus presaging some of the subsequent interpretational difficulties. **Figure 1**, using more than 6,700 isotope data for lead ore deposits from Europe compiled by Hsu in the FLAME database, shows the variation in lead isotope ratios ($^{206}Pb/^{204}Pb$) across Europe. There is a general east-west trend, with lower values in the west ($^{206}Pb/^{204}Pb$ =16.5 – 18.2) and higher in the east (18.6 – 21.4), with the lowest values (13.4 – 16.5) in northwest Scotland and Fennoscandia, and a belt of much higher values ('highly radiogenic', $^{206}Pb/^{204}Pb$ >21.4) between Sweden and Norway. Even though this map for Europe is incomplete, it can be seen why it is not surprising that there is considerable overlap in the isotopic values of metalliferous ores from different sources, at least across mainland Europe and Anatolia.

The scope of the application of lead isotopes to archaeology was vastly increased when researchers realized that it could be applied not only to metallic lead artefacts (which are archaeologically rare) but also to the traces of lead left in silver objects extracted from argentiferous lead ores by cupellation (Barnes *et al.* 1974). Silver is quite common in Europe, and particularly so once metal coinage was introduced around the 8th century BCE in the eastern Mediterranean. However, the utility of the method was even further expanded when it was also applied to the traces of lead in copper objects (Gale and Stos-Gale 1982), present either as a result of smelting impure copper ores, or by some admixture of lead to the copper. The realization that the isotope ratios of lead in archaeological copper objects could be related to differences between specific metal deposits was universally hailed as a major breakthrough in the scientific study of ancient metallurgy. As noted in Chapter 1, traditional provenance studies using trace elements in copper had often only been seen to be partially successful, so the lead isotope method was widely applied to studies of the trade in metals, particularly in Bronze Age Anatolia and the Eastern Mediterranean. Early enthusiasm, however, soon became tempered by differences of opinion in data interpretation, with specific reservations focussing on issues such as the statistical definition of an "ore field" and, given that lead isotope ratio data are not normally distributed, the identification and treatment of "outliers" within the data (Pollard 2009). Modern techniques of data interpretation, such as kernel density modelling (Scaife *et al.* 1996: see below), can now easily overcome such objections, but nevertheless some issues remain. Little consideration has been given to the effect of mixing lead from different isotopic sources, as could happen when copper objects are recycled, or even when lead from different sources is mixed together. The system described here is designed to highlight such processes.

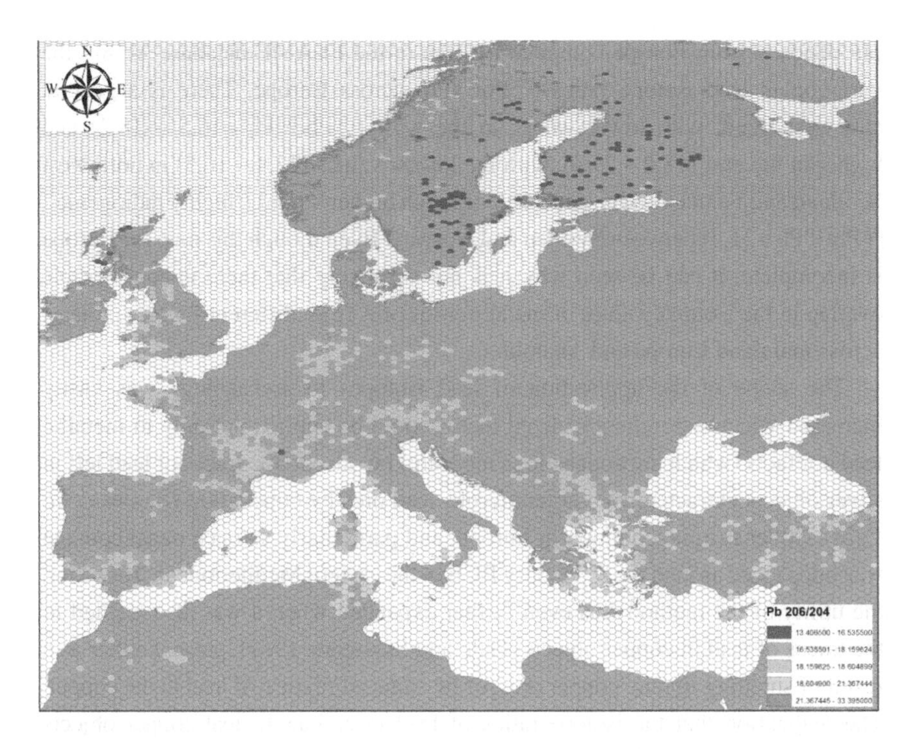

Figure 1:
Variation in $^{206}Pb/^{204}Pb$ in ore deposits across Europe (Hsu: FLAME database)

The evolution of lead isotope ratios: 'ordinary' and 'highly-radiogenic' leads

The conventional means of interpreting lead isotope data in archaeology derives directly from geological practice, which uses pairs of isotope plots (some combination of $^{206}Pb/^{204}Pb$, $^{207}Pb/^{204}Pb$ and $^{208}Pb/^{204}Pb$) to calculate graphically the emplacement age of the lead deposit, based originally on the Holmes-Houtermans equations for the evolution of the isotopic composition of lead deposits (see Pollard *et al.* 2017b, 380–395). At its simplest level, the modern lead isotopic composition of an ore body is controlled by the original isotopic composition of the lead at the time of emplacement, the original amount of uranium and thorium in the deposit, and the geological age of the deposit. In the absence of unusually high concentrations of uranium and thorium, and in a closed geological environment, the isotope ratios of lead in a mineral deposit will evolve along predictable lines over geological time, and the geological age of such deposits can therefore be calculated from a measurement of these ratios (Dickin 2005;

Faure 1986). Such deposits are termed *conformable*, and they tend to have lead isotope ratios close to $^{206}Pb/^{204}Pb \approx 18.5$, $^{207}Pb/^{204}Pb \approx 13.3$, and $^{208}Pb/^{204}Pb \approx 38.3$, which is approximately the average lead isotopic composition in the Earth's crust, as given above. Lead-containing minerals with such ratios are termed *common* or *ordinary* leads. The lead isotope ratios of minerals formed in the presence of excess uranium and/or thorium, however, evolve along different lines as a result of the production of additional quantities of ^{206}Pb, ^{207}Pb, and ^{208}Pb, the exact enrichment depending on time and the amounts of uranium and thorium present. Such deposits are termed *non-conformable*, and their ages cannot be calculated from measurements of the lead isotope ratios using simple evolutionary models—they can give predicted (model) ages which are in the future. They are generally referred to as *anomalous* leads. It is worth pointing out that *all* terrestrial lead deposits contain varying proportions of radiogenic lead, so the use of the term 'radiogenic' in many publications to describe anomalous lead is inappropriate. The term 'highly radiogenic' is more acceptable (e.g., for values of $^{206}Pb/^{204}Pb > 19.5$), and has been used in some of the archaeological literature, and also here.

Geologists were quick to exploit this new chronometric tool based on conformable lead deposits, firstly to obtain an estimate of the age of the Earth, and subsequently to estimate the geological age of the various metalliferous deposits. In practice, however, using lead isotopes to characterize metal deposits is not straightforward. Very few lead deposits are truly conformable, but, conversely, deposits with highly radiogenic lead isotope ratios are relatively rare. Moreover, deposits with highly radiogenic leads can be isotopically very heterogeneous (Liu R. *et al.* 2018). In isotope geochemistry it has now been realised that simple equations such as those of Holmes-Houtermans and subsequent derivatives provide a poor description of the evolution of lead isotopes in terrestrial deposits. This is due to the complexity of the processes acting within the Earth's crust. For example, when an ore fluid is forced through the crust, there can be mixing between the lead in the crustal rock and that in the ore fluid, which can be of quite different ages, resulting in a mineral deposit with isotopic values which do not match either that in the ore fluid nor in the surrounding rock.

Lead isotopes in archaeological objects

During all the archaeological discussion about the use of lead isotopes, little attention was paid to the question of data presentation, in terms of the archaeological utility of using a pair of isotope ratio plots. When we add to the geological complexity described above the possibility of anthropogenic mixing of lead from different sources during smelting and metal production, we might argue that 'conventional' lead isotope bi-plots may not be the best way of looking at the data *from archaeological objects*. If we accept that, at least at certain times and places, the mixing of metal from different sources, or metal recycling, was a significant facet of human behaviour, then the uncritical use of lead isotope data on copper alloys is potentially misleading in terms of provenance (Pollard and Bray 2015). Within FLAME, we have therefore reconsidered the ways in which lead isotopes can be used archaeologically. As before, we focus primarily on assemblages of the objects themselves, and look for *changes* in the isotopic record over time and space. The objective is less to match the objects to specific ores, but more to identify changes in lead isotope values within the hypothesized metal flow, some of which may well be due to changes in ore source, but which also admits other possibilities, such as mixing and recycling. It is obvious that if metal from different sources is being mixed and recycled, then the measured lead isotope signature of the object may not correspond to any one specific source. This is in addition to the equally obvious statement that if lead is added to a low-lead copper base to form a leaded alloy, then the lead isotope signature of the object will be dominated by that of the added lead, not the copper, since the concentration of lead in most smelted copper is low.

In order to explore these ideas, we have proposed a different set of three diagrams which plot the inverse of the lead concentration (1/Pb) in the object against its lead isotope ratio (Pollard and Bray 2015). This parallels the method of presentation used by isotope geochemists for strontium isotope data ($^{87}Sr/^{86}Sr$), with the express purpose of being able to detect mixtures of two or more components. If two different sources of strontium are mixed (e.g., the merging of two rivers, each carrying sediment from different source rock), and each having different chemical abundances and isotopic ratios of strontium, then a plot of 1/Sr vs $^{87}Sr/^{86}Sr$ is used to calculate the strontium concentration and isotopic ratio of the mixture (Faure 1986, 144). Such mixing shows up as a hyperbolic mixing line if the isotope ratio is plotted directly against the strontium concentration, but it becomes linear if plotted as the isotope ratio against inverse concentration (**Figure 2**). The mathematics of the linear relationship between $^{206}Pb/^{204}Pb$ and 1/Pb are shown in the appendix to this chapter. Although for lead there are three such diagrams (1/Pb vs $^{206}Pb/^{204}Pb$, 1/Pb vs $^{207}Pb/^{204}Pb$ and 1/Pb vs $^{208}Pb/^{204}Pb$) rather than one for strontium, this method has proved extremely useful in a wide range of archaeological cases, especially when lead has been deliberately added to the copper alloy.

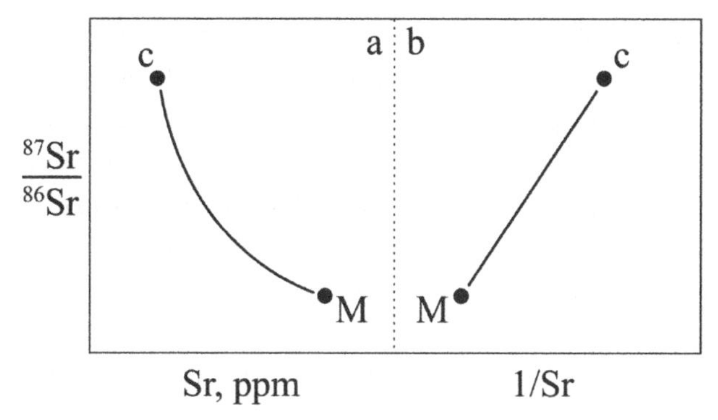

Figure 2:
Isotope mixing lines as seen in strontium isotopes (Faure 1986, 144).

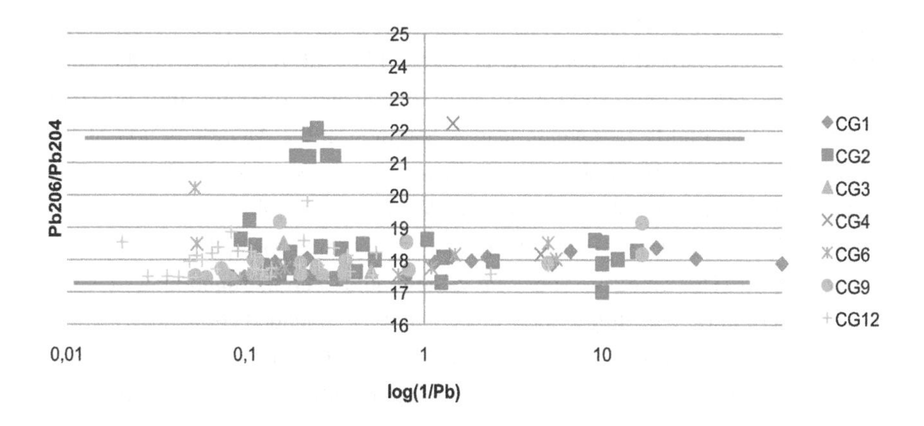

Figure 3:
$^{206}Pb/^{204}Pb$ plotted against 1/Pb for Chinese bronzes from a) the Anyang period of the Shang (c. 1200–1046 BCE) and b) the Western Zhou period (1046–771 BCE) (Liu 2016). The colour coding shows the allocation to Copper Group.

The most significant advantage of this form of presentation is that it provides a way of simultaneously displaying the concentration of lead in an object and its isotope ratio. To illustrate this method, **Figure 3** shows a pair of 1/Pb vs. $^{206}Pb/^{204}Pb$ for Chinese Bronzes in the Sackler Collection, split into the Anyang phase of the late Shang (c. 1200–1046 BCE: Bagley 1987) and the Western Zhou (1046–771 BCE: Rawson 1990). Because the data contain objects with a wide range of lead concentrations (from roughly 0.01% up to 15%), we have presented the horizontal axis as a logarithmic scale, although it is labelled linearly. It is important to note, however, that although this is more convenient for display purposes, it does mean that the linearity of the mixing lines will be reduced.

Because the horizontal axis is inverse concentration, points to the left side of the plot have high lead concentrations, and points to the right are low in lead. Thus 0.1 on the x-axis corresponds to 10% Pb in the original object, and 10 is 0.1%. Although this may not always be true, crudely speaking, we can think of the right hand side of the diagram (with $1/Pb > 1$, or the Pb concentration in the object <1%) as representing the lead isotope values in objects made of unalloyed copper, and the left hand side as being representative of the values in objects with deliberately added lead.

Objects on these diagrams lying on the same *horizontal* line have approximately the same value of $^{206}Pb/^{204}Pb$, and therefore have the potential to come from the same source, or at least from sources sharing similar isotopic values. Thus in these two diagrams we have drawn attention to two horizontal lines—one at approximately $^{206}Pb/^{204}Pb = 17.75$, often referred to in the Chinese lead isotope literature as 'common lead', and one at $^{206}Pb/^{204}Pb \approx 22$, referred to in the same literature as 'highly radiogenic lead'. It is immediately obvious from this pair of diagrams that highly radiogenic lead is common in the Anyang period of the Shang Dynasty, but largely absent in the succeeding Western Zhou Dynasty. The overwhelming majority of data from the Western Zhou fall into a broad band between $17 < {}^{206}Pb/^{204}Pb < 19$, irrespective of whether we look at the data above or below $1/Pb = 1$. This range of $^{206}Pb/^{204}Pb$ values could of course encompass several sources of common lead, but it does indicate that the unleaded coppers ($1/Pb > 1$) have similar isotopic values to that in the added lead ($1/Pb < 1$). Since galenas (PbS) and chalcopyrites ($CuFeS_2$) forming in the same metalliferous veins are likely to have similar lead isotope ratios, this could imply a similar source location for both the lead and the copper. The situation is quite different in the earlier Anyang period data. Here the added lead is distributed between at least two distinct sets of lead isotope values, corresponding to the common and highly radiogenic leads, but the few unleaded coppers represented suggest a definite mixing trend between copper containing 'common' lead ($^{206}Pb/^{204}Pb$ below 20) and the highly radiogenic added lead with a value of $^{206}Pb/^{204}Pb$ around 23. This seems to imply that the sources of copper used in both the Anyang period and the Western Zhou *could* have been the same, but that in the Anyang period it was mixed with highly radiogenic lead, which was largely not used during the Western Zhou.

Several further points can be made about this form of presentation. The first is that the pair of figures shown above are only one of a set of three possible representations, and that, for reasons of space, in Pollard *et al.* (2017a) only this figure was printed (although 1/Pb vs. $^{207}Pb/^{204}Pb$ and 1/Pb vs. $^{208}Pb/^{204}Pb$ were available in the online supplementary material). It is clearly possible that if another lead isotope ratio was used then the diagrams might look different, and a different conclusion could be drawn. It is possible, for example, that significant differences could occur when $^{208}Pb/^{204}Pb$ is plotted instead of $^{206}Pb/^{204}Pb$, since ^{208}Pb is thorogenic (i.e., partially derived by radioactive decay from ^{232}Th), whereas ^{206}Pb and ^{207}Pb are both uranogenic (from ^{238}U and ^{235}U, respectively). Geochemically thorium is more insoluble than uranium, and is therefore less widely dispersed in the Earth's crust than uranium. The plots of 1/Pb vs. $^{207}Pb/^{204}Pb$ and 1/Pb vs. $^{208}Pb/^{204}Pb$ for the same data as Figure 3a are shown here as **Figure 4** and in this case show no appreciable difference. In our experience this is usually the case, but it will not always be so, so all three representations should be studied before any conclusions are reached.

a)

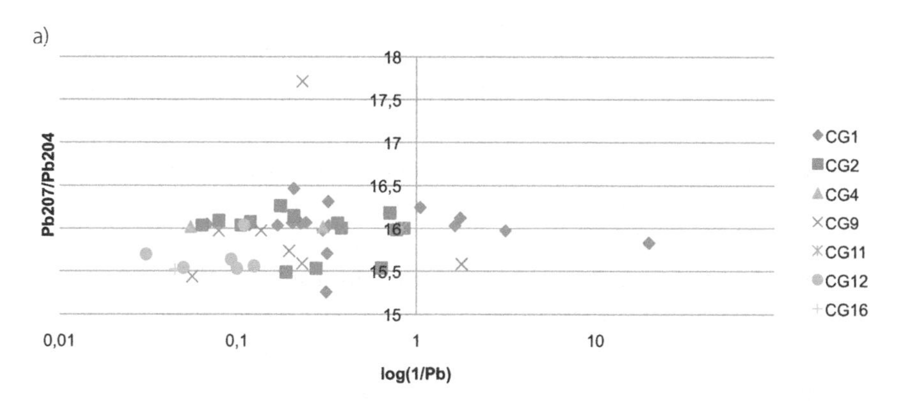

b)

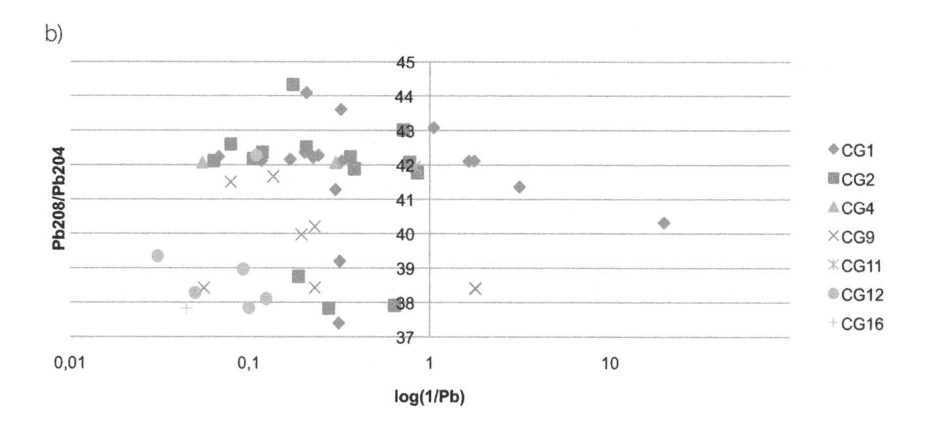

Figure 4:
a) $^{207}Pb/^{204}Pb$ vs 1/Pb and b) $^{208}Pb/^{204}Pb$ vs 1/Pb for same data as shown in Fig. 3a.

Another point to be made is that, as shown in these figures, we can also add the results of the trace element categorisation (Copper Groups) by colour coding the points, enabling us effectively to combine lead isotope, trace element and alloying information in a single diagram. In Figure 3a, for example (the Anyang Shang data), there appears to be a difference in the isotopic values of the lead associated with different copper groups. CG1 (Cu only, with no trace elements above 0.1%), for example, seems to be preferentially associated with Pb having a highly radiogenic signal, which may be useful in considering the sources of the copper and highly radiogenic lead. Conversely, CG12 (Cu with As, Sb and Ag) seems to be primarily alloyed with common lead—again, this is likely to have some significance in considering sources.

One significant observation is that, although modern lead isotope ratios are routinely measured to four or five decimal places, the lead isotope data as plotted in these figures are only discussed at the level of the first or second decimal place. This might indeed be wasteful in terms of information content, but when dealing with archaeological objects which may contain mixed or recycled lead, or even just considering the natural variation within ore sources, it is likely that in some cases only the first two decimal places are archaeologically significant. There may well be particular datasets where the full precision of the measurements might be needed, even when using this method of presentation, but just because we can measure an isotope ratio to five decimal places does not necessarily mean that it has archaeological significance to five decimal places.

We have also experimented with plotting the lead isotopes against elements other than lead. Although such plots lack the theoretical underpinning of a 1/Pb vs lead isotope ratio plot, they can nevertheless prove informative about mixing processes. Since nickel is resistant to oxidative loss, we can plot nickel as 1/Ni against the lead isotope ratio and use the absolute level of nickel as an indicator of dilution (since the most likely way of decreasing the nickel content of a copper alloy is to mix it with metal which contains no nickel; i.e., by dilution). If we see a change in lead isotope value as a function of the nickel content, we can suggest that there has been a mixing process between two units of metal with different lead isotope values, but one containing higher nickel and one with little or no nickel. **Figure 5** shows a comparison of 1/Ni vs. $^{206}Pb/^{204}Pb$ for Shang Anyang and Western Zhou bronzes from the Sackler collection, to be compared with Fig. 3 for lead. There is again a clear difference between the Shang and Western Zhou datasets. The Shang data shows samples with much lower nickel content (higher 1/Ni), and with clear mixing lines between the lower and higher values of nickel, but starting at the highly radiogenic value of $^{206}Pb/^{204}Pb$ and joining to both the common and highly radiogenic lead values. This is complicated to interpret since the nickel is most likely to be associated with the copper, and the higher values of lead are likely to be separately added, but it could indicate a mixing between the common and highly radiogenic sources of lead. This utility of such approaches requires further investigation.

a)

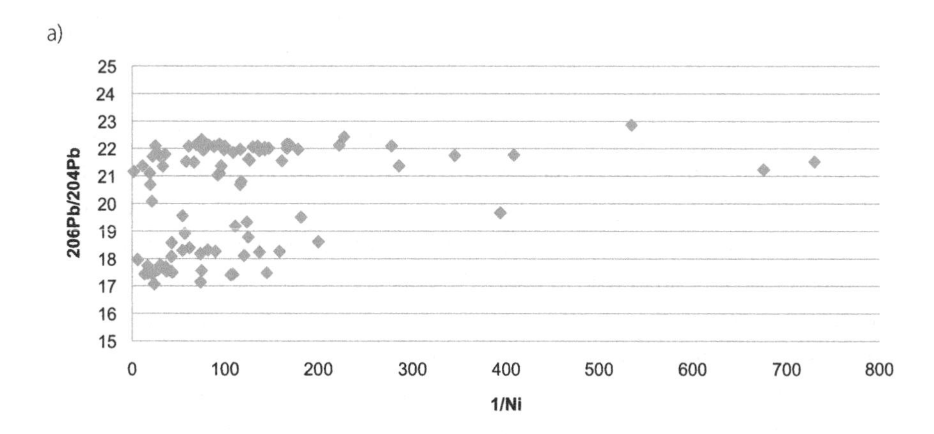

b)

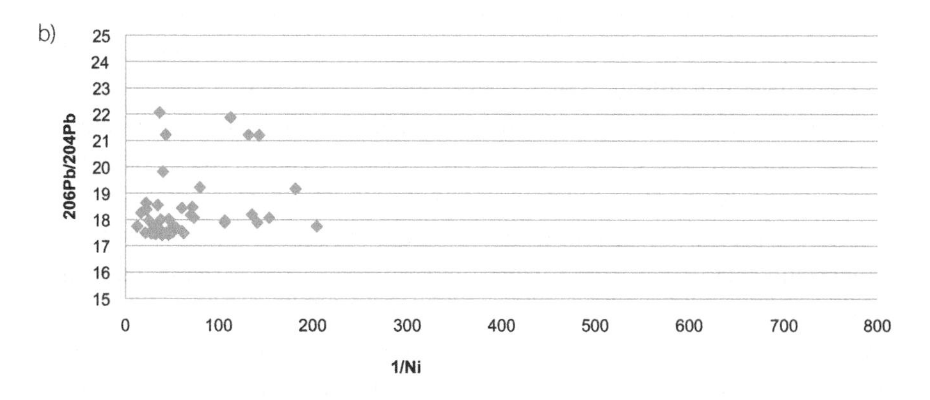

Figure 5:
$^{206}Pb/^{204}Pb$ plotted against 1/Ni for Chinese bronzes from a) the Anyang period of the Shang (c. 1200–1046 BCE) and b) the Western Zhou period (1046–771 BCE).

Changes in lead isotopes in assemblages over time

Although for some regions there is now a relatively substantial number of lead isotope measurements on Bronze Age archaeological metal artefacts in the database, unfortunately, the majority of these are not associated with chemical data from the same object. This is because labs which are capable of measuring lead isotopes are often not equipped with instruments for chemical analysis, and *vice versa*. Without a measurement of the lead content in the object, we cannot produce 1/Pb vs isotope ratio plots for individual objects. We can, however, produce something almost as useful by simply plotting one of the measured isotope ratios in a chronological sequence of samples. This works well if we have a block of isotope data for each of several consecutive periods of time, and plot each period in the correct order. By grouping the data according to the chronological sequence we can reveal any patterns of change in the isotope values over time, which might be indicative of changing ore sources or reveal periods of intensive mixing. Within each group, however, the order of the samples along the horizontal axis is arbitrary. The method works equally well for comparing isotopic data from different contemporary sites. Objects with similar isotopic ratios will still show up as strong horizontal lines in these diagrams. If we make the assumption that horizontal lines with different isotopic values represent different sources (although single sources can of course show a range of values), then changes in the value of these lines *may* represent a switch in ore source. Essentially, these diagrams combine in a very simple way both isotopic and archaeological information.

Here we present the data in terms of $^{206}Pb/^{204}Pb$, but, as with the 1/Pb vs isotopic ratio diagrams discussed above, similar figures can be produced for both $^{207}Pb/^{204}Pb$ and $^{208}Pb/^{204}Pb$, which again show nothing different to the figure shown here. **Figure 6** illustrates the method using lead isotope data in vessels from Bronze Age central China (Jin *et al*. 2017). In fact this figure shows the combination of a chronological sequence and a comparison of objects from contemporary sites. The chronological sequence consists of Erlitou (c. 1900–c. 1500 BCE), the Erligang (Zhengzhou) period of the Shang (c. 1500–c. 1400 BCE), the Anyang period of the Shang (c. 1400–1046 BCE) divided into the four Yinxu phases as discussed in Chapter 5, and the Western Zhou period (1046–771 BCE). The Erlitou period is represented only by samples from the Erlitou site itself, whereas the Erligang period has samples from Zhengzhou (the Erligang capital) and also from the Erligang period sites of Yuanqu and Panlongcheng. The samples representing the Anyang period all come from excavations at Anyang. The Western Zhou samples are from the Sackler collection (Bagley 1987), and are largely unprovenanced.

The most distinctive feature, as discussed above in relation to Figure 3, is the presence of so-called highly radiogenic lead from the Erligang period through to the late Anyang period. In the preceding Erlitou period there is no evidence for the use of this highly radiogenic lead (taken here to be $^{206}Pb/^{204}Pb$ above *c.* 19), but there seems to be at least two distinct common isotope values present at Erlitou ($^{206}Pb/^{204}Pb$ ~ 16.5–17 and $^{206}Pb/^{204}Pb$ ~ 18–18.5), with possibly some mixing between them. The lower of these two signatures ($^{206}Pb/^{204}Pb$ ~ 16.5) appears to continue into the Erligang (Zhengzhou) period (*c.* 1500–1400 BCE), but the higher one is less well represented, if at all, in Erligang. This common lead is, however, supplemented during the Erligang period by lead containing a highly radiogenic component ($^{206}Pb/^{204}Pb$ ~ 19-23), the significance of which is discussed further below. This Erligang pattern is broadly continued into the Anyang period of the later Shang (*c.* 1400–1046 BCE), but with some differences in detail. The earliest phase, Yinxu I, has a wide scatter of lead isotope values, ranging from $^{206}Pb/^{204}Pb$ below 16, up to highly radiogenic values of 23.5. In this respect it matches more closely the previous Erligang (Zhengzhou) and Panlongcheng patterns than it does the subsequent Yinxu phases. Yinxu Phase II is scattered across a similar but not identical range, consisting of a tight 'low' grouping at $^{206}Pb/^{204}Pb$ ~ 18-19, which could correspond to the higher of the two earlier Erlitou groups, and going up to radiogenic values of > 24. Phase III shows a highly radiogenic group between 19 and 22, and predominantly (but not exclusively) the same common group at around 18 as seen in Phase II. However, Phase IV is strikingly different, in that, with the exception of two points, the highly radiogenic lead has disappeared and the values are mostly very consistent around a common value of 17.5, which is similar to the common lead values seen most notably at Panlongcheng. The Western Zhou data include a continuation of this 'low' source (~ 17.5) which first appears in Yinxu Phase III and dominates Phase IV, but also contains a wider scatter of common lead with values between 18-19, which could include the common lead source identified in Yinxu Phase III. In the Western Zhou data, the highly radiogenic component has also virtually disappeared. The source of this highly radiogenic lead is one of the major questions in Chinese archaeometallurgy, and, although this method does not enlighten us as to where it comes from, it does show that it was accessed suddenly with the move of the Shang capital to Erligang (Zhengzhou), and ceased before the fall of Anyang, probably during the Yinxu III phase.

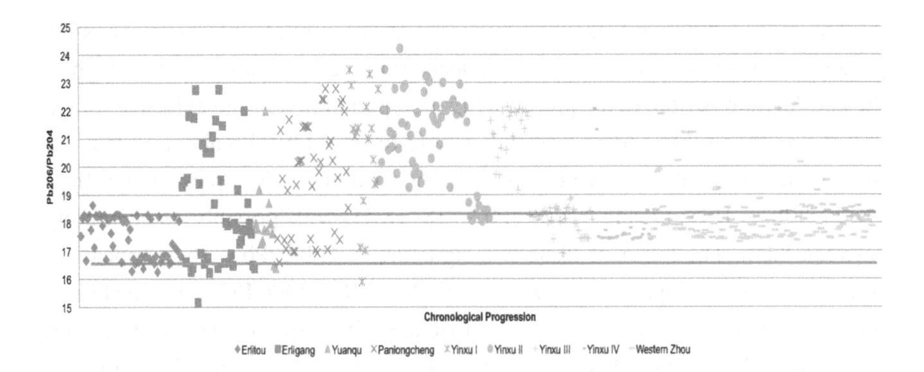

Figure 6:
Values of $^{206}Pb/^{204}Pb$ for archaeological objects from Central China from the Erlitou culture to Western Zhou (Jin ***et al.*** 2017). (Data from Jin ***et al.*** 2001; Tian 2013; Peng ***et al.*** 1999, 2001; Sun ***et al.*** 2001; Rawson 1990)

Lead isotope measurements have made a major impact on archaeological research over the last thirty years, but little thought has been given to presenting the data in such a way as to link it directly to archaeological questions. When presented, as here, in a different way, and when combined with chemical and other evidence, lead isotopes can provide an extremely powerful and independent tool for understanding the circulation of metal. As in the two previous chapters, however, in order to fully exploit this, this approach has to be linked with a shift in the emphasis of the question away from one of simple 'provenance' towards a more complex picture, of which 'provenance' is but one component.

The source(s) of Chinese highly radiogenic lead: geochemical parameters (Mu-Kappa-T)

More than thirty years have passed since the first report of the finding of highly radiogenic lead in Shang Dynasty Chinese bronzes (Jin 1987). Since then, the use of highly radiogenic lead in Bronze Age China has been frequently mentioned and discussed in many important publications. The most recent and high profile debate was triggered by Sun *et al.* (2016), who argued that the metal contained in the ritual bronze vessels unearthed at Anyang originated in Africa. Although extremely eye-catching, this paper has subsequently been heavily criticised, especially on archaeological grounds (Liu S.R. *et al.* 2018). Figure 6 above shows the widespread use of highly radiogenic lead in Chinese ritual bronzes from the Erligang (Zhengzhou) period (ca. 1500–1300 BC) until the Yinxu II Period of

Anyang (ca. 1200–1046 BC). Even more remarkably, isotopic measurements of bronzes from outside the Central Plains of China—that is, from Sanxingdui in Sichuan, from Panlongcheng and Xin'gan on the Middle Yangzi River—also show the use of highly radiogenic lead in their cast bronzes. Equally intriguing and mysterious, highly radiogenic lead appears to have disappeared in all areas by the subsequent Zhou dynasty. This provides one of the most enduring problems in Chinese archaeology, and a great challenge to lead isotope geochemists—where does the radiogenic lead come from, and why does its use appear to rise and fall synchronously over a large area in relatively short time period?

Another way of looking at this question is to ask 'why hasn't lead isotope geochemistry already solved this issue?' One of the long-standing criticisms of the use of lead isotope data in archaeometallurgy has been the tendency to use lead isotope ratios as simple *numbers* which can potentially be used to characterise an ore source (e.g., Albarède *et al.* 2012). According to these commentaries, this ignores a crucial fact about lead isotopes—they evolve according to well-understood laws of radioactive decay and geochemistry. In other words, these numbers can tell us something about the geochemical and geochronological characteristics of the parent ores. An alternative approach has been described by Albarède *et al.* (2012), which uses the measured isotope ratios for each sample to calculate the parameters μ (the ratio of $^{238}U/^{204}Pb$ in the parent deposit), κ (the ratio of $^{232}Th/^{238}U$ in the deposit), and T (the 'model age' of the deposit). These have the advantage of being three independent geological parameters related to the characteristics of the ore source. The model age T is an estimate of the geological age of the province in which the ores are found. Because of the assumptions invoked by the use of simple models for the emplacement of the ore deposit, the values of T do not necessarily correspond to real geological ages (indeed, sometimes the model ages predict the formation of the deposit in the future), but they are usually good enough to identify geological provinces by their tectonic formation ages. Nevertheless, these three parameters describe the geochemical environment of the parent deposit, which should be similar within a particular deposit, but different between deposits of different ages and chemistry. Albarède *et al.* (2012) use the equations for the primary growth of radiogenic ^{206}Pb, ^{207}Pb and ^{208}Pb from ^{238}U, ^{235}U and ^{232}Th respectively (see Pollard *et al.* 2017b, 388–390) to generate a set of three simultaneous equations in terms of T, $\Delta\mu$ and $\Delta\kappa$, where $\Delta\mu = \mu - \mu^*$ and $\Delta\kappa = \kappa - \kappa^*$, and the asterisk signifies the value of each parameter in modern common lead ($\mu^* = 9.66$ and $\kappa^* = 3.90$). In this formulation, we can solve the three equations for three unknowns T, $\Delta\mu$ and $\Delta\kappa$ using a trust-region-dogleg algorithm to ensure rapid convergence, and then μ and κ are easily obtained from the expressions for $\Delta\mu$ and $\Delta\kappa$. The model age can (usually) be used to assign the origin of ores to a particular geological province, and variations in μ and κ can help to discriminate segments within geological provinces.

We have used calculations of T, μ, and κ to further explore the variations within the highly radiogenic lead used in the Shang Dynasty of China. In most Shang and Zhou ritual bronzes, the lead levels are sufficiently high (~5-20%) that the measured lead isotope ratio in a particular object must reflect that in the added lead, rather than the traces introduced by the copper or tin. During the 1980s, Jin Zhengyao and his colleagues argued that the lead contained in the ritual vessels found at Anyang might have originated in southwest China, specifically in northeast Yunnan (Jin 1987; Jin *et al.* 2004), since this appeared to be the only region in China capable of yielding such highly radiogenic lead. Furthermore, the predominance of objects containing highly radiogenic lead at the sites of Sanxingdui (Jin *et al.* 1995), Hanzhong (Wang *et al.* 2008) and Jinsha (Jin *et al.* 2004) in southwest China, and Panlongcheng and Xin'gan (Jin *et al.* 1994) in the south, seemed to reinforce the idea of a lead supply from somewhere south of the Yangzi river. This suggestion, however, raised some difficult questions for specialists in Chinese bronzes. Yunnan is very distant from Anyang, especially since there are many other sources of lead closer to the Central Plains, and there is as yet insufficient archaeological evidence to understand the nature of contact between the Central Plains and Yunnan in the Shang dynasty. There are also other, equally difficult, questions to consider. The phenomenon of the use of highly radiogenic lead in Chinese ritual bronzes of the Bronze Age is relatively short-lived, as discussed above. It 'switches on' at the beginning of the Erligang phase of the Shang (c. 1500 BCE), and is virtually gone by the end of the Anyang phase, around 1046 BCE, or even before (the end of Yinxu III). During these c. 400 years, the use of highly radiogenic lead is ubiquitous in the Shang world and beyond. Intuitively one feels that this must represent the widespread use of a single source of lead—otherwise how would smelters over a large area simultaneously switch to, and then switch away from, the use of highly radiogenic lead, assuming that they had no way of distinguishing between highly radiogenic and common lead? Conversely, if it is a single source, then how was the distribution system organised so that it simultaneously became ubiquitous at the Shang capital, but also at many other largely autonomous centres such as Sanxingdui in Sichuan Province?

In order to see if the available data on highly radiogenic lead in Bronze Age China suggests a single or multiple source, we have assembled all the available isotopic analyses used in Figure 6 from objects containing highly radiogenic lead (arbitrarily defined as those having values of $^{206}Pb/^{204}Pb > 19$), and converted the data to T, μ, and κ.

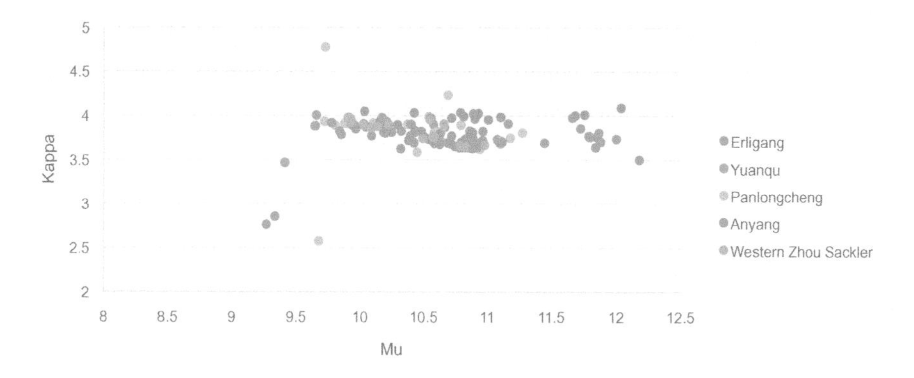

Figure 7:
Highly radiogenic lead data ($^{206}Pb/^{204}Pb > 19$) from Fig. 6 converted into T, μ, and κ.

Figure 7 shows all the available lead isotope data for highly radiogenic lead (as defined above) for archaeological objects from the three Erligang period sites of Erligang, Yuanqu and Panlongcheng and the later Shang period at Anyang, plus the data from Bagley (1987) on the Western Zhou objects in the Sackler Collection, plotted as κ ($^{232}Th/^{238}U$) vs. μ ($^{238}U/^{204}Pb$). A striking feature is that the vast majority of data all appear to overlap in one group elongated along the μ axis, which is indicative of highly variable initial concentrations of uranium. However, the total range of this group is surprisingly large, having values of μ between 9.77 and 11.16, and κ between 3.67 and 4.09. Typical ranges for common sources of lead are much narrower, such as 9.64 – 9.70 for μ and 3.83 – 3.86 for κ in Mexican ores (Cumming *et al.* 1979). One would of course expect a source of highly radiogenic lead to have a larger range of values than a common source. The key question is do these data represent a single source, or more than one source? Looking more closely at Figure 7, there is the possibility that the Anyang data are split into two groups, one as just described (heavily overlapped by the other sites) and another with μ between 11.66 and 12.18. This could simply be an artefact of the small sample numbers, but the fact that it is *only* data from Anyang which occupies this space suggests that there may have been more than one source of highly radiogenic lead used at Anyang.

In an attempt to match these radiogenic data to a specific area, we have created an admittedly incomplete set of data on modern highly radiogenic Chinese lead ores, which, after conversion to μ, κ and T, shows a total range for μ of approximately 10 – 20, and 2.5 – 6 for κ (**Figure 8**). Plotting μ and κ for highly radiogenic lead by province, we find two substantial matches for the highly radiogenic leads found in Shang bronzes (the range of which, taken from Fig. 7, is shown by the blue box), one from Shandong and one from Yunnan province.

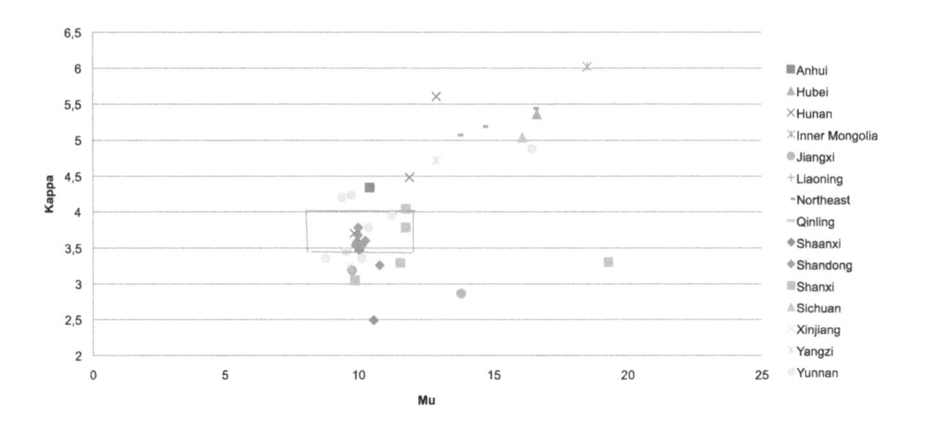

Figure 8:
Highly radiogenic lead ($^{206}Pb/^{204}Pb > 19$) in modern ore data from China, converted into T, μ, and κ.

This plotting of data from modern Chinese lead ores reveals some interesting but as yet far from conclusive suggestions as to where the radiogenic lead might have originated. Jin *et al.* (2017) observe that the Zhongtiao mountains (close to Erlitou and Zhengzhou) have some evidence for highly radiogenic lead (Xu *et al.* 2005), albeit only so far published as measured in chalcopyrite samples. Similarly, the Qinling mountains to the west of the central plains have some evidence for highly radiogenic lead (Zhu *et al.* 2006). As originally noted by Jin, northern Yunnan in the south-west of China remains a possible source of this metal (Jin *et al.* 1994; 2004).

Plotting the mu-kappa data for highly radiogenic leads found in Shang dynasty bronzes has shown that those from Erlitou, Zhengzhou, Panlongcheng and Yuanqu all occupy the same extended region of the graph, which could indicate that they come from a single source. There is some evidence that there may be a second source represented at Anyang, but not at the other sites. As yet the modern ore data are too limited to consider asking where these sources might be, but the original suggestion of Yunnan appears to be a possibility, alongside Shandong, although other, closer, sources such as Zhongtiao Mountain must also be considered.

All of this helps to shed more light on the phenomenon that is the use of highly radiogenic lead in Shang dynasty China, but does not specifically identify the number of sources, nor their locations. Two other pieces of the jigsaw have recently been put together, both of which point (independently) to the use of multiple sources of highly radiogenic lead. One is the study of silver levels in the bronzes. As noted in Chapter 3, silver is unlike the other trace elements used to create Copper Groups because in leaded bronzes it is possible that it accompanies the lead and not the copper. In a recent study, Pollard *et al.* (submitted) concluded that there were at least two patterns of association between the silver and the lead, indicating two different geological provinces for the origin of the lead. Interestingly, there did not appear to be a clear difference between the Ag-Pb association in highly radiogenic and common leads where they co-occurred, suggesting that the two types of lead may have come from similar geological environments. Perhaps even more significantly, Liu R. *et al.* (2018) have extended the data shown in Figure 6 above to include lead isotope ratios from glasses, glazes and pigments into the Han and Tang dynasties. This shows clearly that highly radiogenic lead continued to be used for a wide range of materials and across a broad area of China well into the Tang dynasty. This suggests that sources containing highly radiogenic lead were probably widely available across China, thereby strongly supporting the likelihood that multiple sources of highly radiogenic lead *could* have been used by the Shang and their neighbours.

Kernel Density Estimates of lead isotope distributions

The practice that developed in the 1990s of drawing '90% confidence ellipses' around archaeological data presented in an isotope ratio bi-plot to define an ore field has drawn much criticism. It either assumes a normal distribution of points in the space defined by the isotope ratios, which has been shown not to be true (Baxter and Gale 1998), or is an entirely arbitrary exercise, which often involves dropping data points as 'outliers' in order to achieve a regular ellipse. The use of kernel density estimates (KDEs) has been shown to be much more robust in defining the actual distribution of an ore field (Baxter *et al.* 1997; Scaife *et al.* 1999), but has not been widely adopted until recently because of the lack of simple algorithms to perform the calculation.

We now use the software developed by Christopher Bronk Ramsey for producing and comparing KDE distributions from the chemistry of volcanic tephra, and available within the open access database developed for the RESET project in Oxford (*Response of Humans to Abrupt Environmental Transitions*: c14.arch.ox.ac.uk) (Bronk Ramsey *et al.* 2015). As introduced in Chapter 5, kernel density estimates are a non-parametric way to convert continuous data into a smoothed probability density function, using a smoothing parameter h, called the kernel. Using KDEs has three main advantages. Firstly, KDEs do not assume that the data has to be normally distributed, which is particularly useful for lead isotope ratios (although the way the kernel h is usually derived is to use an approximation which becomes more optimal the closer the distribution is to normality). Secondly, KDEs can produce smoother distributions than conventional histograms whose appearance is significantly affected by the choices of bin width and start/end points of bins (see Chapter 4). Thirdly, the KDEs can easily generate a multi-dimensional visualisation to compare different datasets in a way that histograms cannot. In archaeology, KDEs have been found useful to interpret chemical compositions, lead isotope ratios and spatial data (Baxter *et al.* 1997). Here we show that kernel density distributions from different assemblages can also be compared to estimate the likelihood of overlap—i.e., to answer the question 'are these distributions the same or different'.

The univariate kernel density estimator ($\hat{f}_h$) of the parent function (f) from which the samples are derived is mathematically defined as follows:

$$\hat{f}_h(x) = \frac{1}{n}\sum_{i=1}^{n} K_h\left(x - x_i\right) = \frac{1}{nh}\sum_{i=1}^{n} K\left(\frac{x - x_i}{h}\right)$$

where n is the sample size, h is the kernel bandwidth and the kernel K is a summed probability density function over all the samples. The choice of kernel bandwidth h is crucial for the shape of the KDE and the subsequent interpretations. If h is too wide the distribution will be over-smoothed and may possibly hide important detail, but if it is too narrow the distribution will be spikey and possibly discontinuous. Many mathematical formulations have been put forward to select optimal kernel widths in univariate cases, but there is no well-defined method to establish the optimal kernel bandwidth in multidimensional data. According to Bronk Ramsey *et al.* (2015), a commonly-used approximation for the optimal bandwidth of the multidimensional kernel is:

$$h = \left\{\frac{4}{(d+2)n}\right\}^{\frac{1}{(d+4)}}$$

where d is the number of dimensions (parameters measured) and n is the number of samples in the dataset. This estimate approaches optimality as the data approach multivariate normality. The details of the derivations are given in Bronk Ramsey *et al.* (2015, 42).

Apart from producing smoothed quasi-continuous probability distributions for univariate or multivariate distributions, a key additional capability is to test the degree of overlap between two distributions. In the original application of this approach to tephrochronology, the purpose was to answer the question, based on a set of chemical measurements, does the volcanic glass found in place A match that of a particular source volcano? This methodology is, however, equally applicable to the chemical composition of metal assemblages, and also to lead isotope ratios. Again, Bronk Ramsey *et al.* (2015) provide the detailed mathematical calculations for comparing the chemical or isotopic compositions of two assemblages.

If we have two assemblages *X* and *Y* containing *n* and *m* samples respectively, then we can use the kernel density estimates derived above for *X* and *Y* to answer the question 'how likely is it that each of the individual samples in assemblage *Y* could belong to the distribution defined by *X*? Mathematically the average value for this probability is given by:

$$q_x(x) = q_x(x_1, \ldots\ldots x_d) = \frac{1}{n}\sum_{i=1}^{n} N_d\left(x_i, h^2 \sum x\right)$$

$$q_{x_j}(x) = \frac{1}{n-1}\sum_{i=1, i\neq j}^{n} N_d\left(x_i, h^2 \sum x\right)$$

This absolute value of the average is not very useful because it depends on how the various parameters are standardized. It is better to normalise it against the average probability of the individual values of the *n* samples in *X* belonging to *X*. This gives a ratio of the two probabilities:

$$B_{yx} = \frac{\frac{1}{m}\sum_{i=1}^{n} q_x(y_i)}{\frac{1}{n}\sum_{j=1}^{n} q_{x_j}(x_j)}$$

$$B_{xx} \equiv 1$$

This ratio (B_{yx}) is a measure of the likelihood of assemblage *Y* being a sub-set of assemblage *X* (which is not the same as B_{xy}, the likelihood of *X* being a sub-set of *Y*). This value should approach 1 if the distribution of the two datasets is similar, but could exceed 1 if dataset *Y* is located on top of the densest area of dataset *X*. Moreover, determining the threshold for the significance of matches is arbitrary. It is strictly not possible to demonstrate consistency between two datasets, only dissimilarity. Thus if the value of the ratio falls below 0.05 then it is probably safe to say at the equivalent of 95% confidence that the two distributions are different. It is more difficult to be certain about when we can accept the likelihood of similarity between two assemblages, but clearly values close to or greater than 1 would mean that we cannot disprove the possibility of them being from the same source.

Figure 9a shows two conventional sets of lead isotope diagrams ($^{207}Pb/^{204}Pb$ vs. $^{206}Pb/^{204}Pb$ and $^{208}Pb/^{204}Pb$ vs. $^{206}Pb/^{204}Pb$) for data from north-west China (Hsu *et al.* 2018). The sites are the Wangdahu cemetery in Ningxia (WDH), modern galena ore data from the Dajing mine in Inner Mongolia (DAJ: Chu *et al.* 2002), plus data from metallurgical remains found at smelting sites associated with the Upper Xiajiadian culture (UX_ORE and UX_SLAG, for ore and slag samples respectively, from Dong 2012), and thought to derive at least in part from the Dajing mine (Yang 2015, 183). A striking feature of these diagrams is the linear array of isotope data for both uranium- and thorium-derived Pb for both Wangdahu and Dajing, but with different gradients. It is clear from the raw data that the lead in the objects recovered from the Wangdahu cemetery did not originate from the Dajing mine. However, some points of the Upper Xiajiadian ore and slag analyses seem to fall exactly on the straight line derived from the Dajing ore data, whereas other samples scatter to both sides of this line. Using KDE (**Figure 9b**), we can better visualise the data, and also statistically test the probability of association between the various assemblages represented.

The overlap between all pairs of probability distributions can be expressed as a matrix based on KDE estimates as described above (**Table 1**), using all three ratios (i.e., calculating the three dimensional overlap). It is clear that the Upper Xiajiadian ore and slags are substantially overlapping (UX_ORE and UX_SLAG = 1.137, UX_SLAG and UX_ORE = 0.907; see discussion above about the reason for this non-commutativity), and are both very likely to be a subset of the Dajing ore data given that the probabilities are very close to 1 (UX_ORE and DAJ, UX_SLAG and DAJ $\approx$ 0.95). There is no overlap between the Wangdahu data and any of the other data in the analysis. This makes sense since Dong (2012) has already argued that Xiajiadian people transferred Dajing ores to other places for further smelting whereas, as noted above, the WDH bronzes are highly unlikely to have been made from the Dajing ores or recycled Xiajiadian objects.

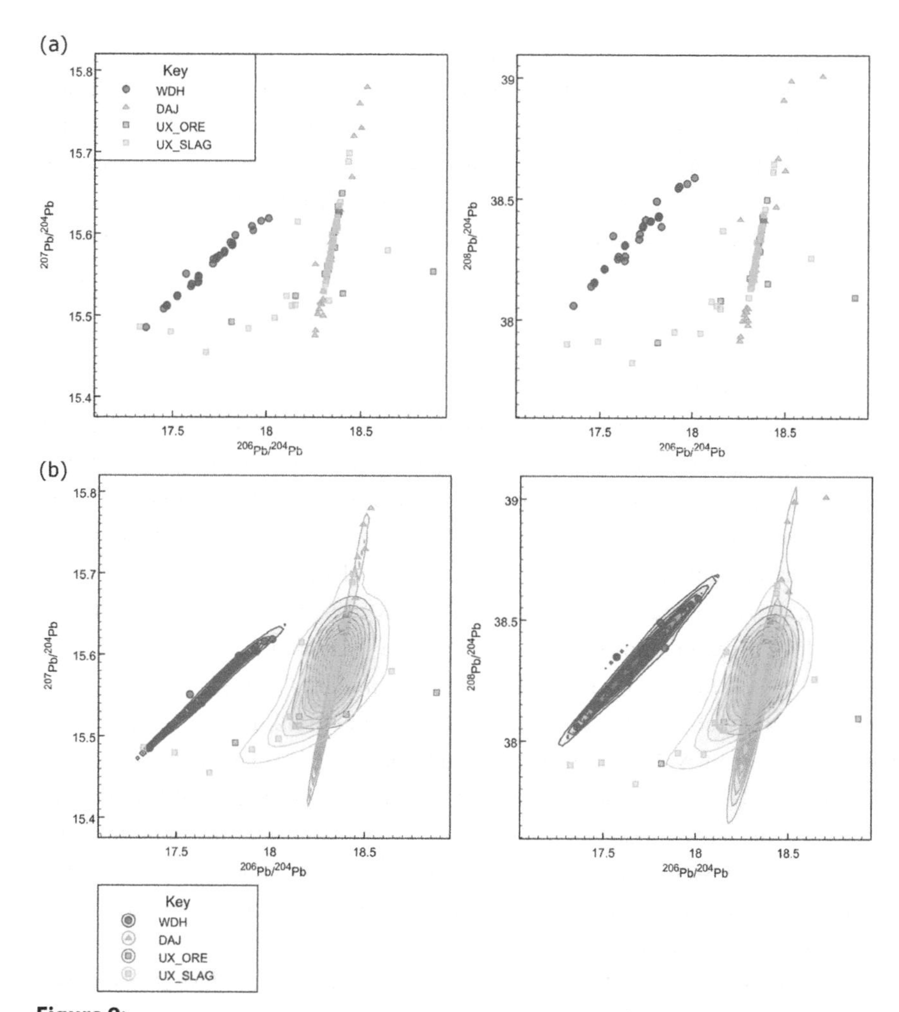

Figure 9:
a) Conventional isotope bi-plots of lead isotope data from north-west China, and b) a KDE rendition of the same data (Hsu ***et al.*** 2018).

	KDE probability distributions ($^{206}Pb/^{204}Pb$, $^{207}Pb/^{204}Pb$, $^{208}Pb/^{204}Pb$)			
Datasets	A - WDH	B - DAJ	C - UX_ORE	D - UX_SLAG
A - WDH	1	0	0	0.01564
B - DAJ	0	1	**0.552**	**0.81104**
C - UX_ORE	0	**0.94657**	1	**1.13702**
D - UX_SLAG	0	**0.94983**	**0.90702**	1

Table 1:
Overlap of KDEs for the various groups shown in Fig. 9 (Hsu ***et al.*** 2018)

Although it is helpful to see these results expressed quantitatively, we should note that the numerical value of ≈ 0.95 for the overlap between either UX-ORE and UX-SLAG and the Dajing data overlooks the fact that both sets of UX data extend far beyond the narrow line of the Dajing data. This asymmetric relationship is confirmed by the lower overlap between DAJ and UX-ORE (0.55) and DAJ and UX-SLAG (0.81), which are both lower than the reverse overlaps (0.95). In other words, some of the UX ore samples do not match exactly the Dajing ore data, suggesting that either the current Dajing data is not completely representative of the Dajing ores, or that ores from other areas were also smelted at these Upper Xiajiadian smelting sites.

KDEs provide the best rendition of the distribution of lead isotopes in an assemblage, and are clearly superior to distributions using 'confidence elipses', which are based on an assumption of normality in all three isotope ratios. As shown above, they can also be used to estimate the similarity (strictly, dissimilarity) between two assemblages by calculating the degree of overlap between the two probability distributions. This is an extremely useful quantitative method for assessing the similarity between the lead isotope data from different assemblages. It can also be used to calculate the similarity between groups defined by multivariate chemical data, as shown for tephra (Bronk Ramsey *et al.* 2015), and also for medieval glass (Bidegaray and Pollard 2018). The previous chapter showed that KDEs can also be used to represent the distribution of the alloying elements (or, indeed, trace elements) in an assemblage, and, by extension, can be used to compare distributions between assemblages. Potentially this could therefore provide an alternative to the use of cumulative frequency distributions (CFDs) and comparison using the Kolmogorov-Smirnov test.

The examples provided in this chapter have been chosen to illustrate a number of different ways in which lead isotope data can be used on archaeological material. Conventional lead isotope bi-plots, adopted from isotope geochemistry, have proved useful in some circumstances, but do not directly address archaeological situations where there is the potential for anthropogenic mixing of lead in archaeological objects. As with our consideration of trace elements and alloying elements, the methods presented here have been derived from a conviction that it is detecting *change* in the archaeological record which is significant, rather than a more limited focus on the ideas of provenance.

Appendix: Mixing model for lead isotopes

Assume that we create a mixture of two components A and B in differing proportions, specified by a mixing parameter f_A defined as:

$$f_A = A/(A+B) \tag{1}$$

where A and B are the weights of the two components in the mixture. If each component contains two elements (e.g., Cu and Pb), and Pb_A and Pb_B are the concentrations of Pb in components A and B, expressed in weight units, then the concentration of Pb in the resulting mixture is

$$Pb_M = Pb_A \cdot f_A + Pb_B \cdot (1 - f_A) \tag{2}$$

Rearranging this expression shows that Pb_M, the weight of Pb in the mixture, is a linear function of f_A:

$$Pb_M = (Pb_A - Pb_B)\, f_A \cdot + Pb_B \tag{3}$$

Consider now that the two components A and B not only have different concentrations of lead but also different values of the set of three lead isotope ratios, having ^{204}Pb as denominator ($^{206}Pb/^{204}Pb$, $^{207}Pb/^{204}Pb$, $^{208}Pb/^{204}Pb$). If the ratio of $^{206}Pb/^{204}Pb$ in component A is $(^{206}Pb/^{204}Pb)_A$, and that in B is $(^{206}Pb/^{204}Pb)_B$, then the ratio in the mixture $(^{206}Pb/^{204}Pb)_M$ is given by:

$$(^{206}Pb/^{204}Pb)_M = (^{206}Pb/^{204}Pb)_A\, (Pb_A/Pb_M)\, f_A + (^{206}Pb/^{204}Pb)_B\, (Pb_B/Pb_M)\, (1 - f_A) \tag{4}$$

This equation contains two different weighting factors: – f_A which relates to the abundances of the two components A and B, and Pb_A/Pb_M and Pb_B/Pb_M, which relate to the fractions of Pb in the mixture contributed by A and B. We can eliminate f_A from this equation by rearranging eqn. (3):

$$f_A = (Pb_M - Pb_B)/\, (Pb_A - Pb_B) \tag{5}$$

Giving:

$$({}^{206}Pb/{}^{204}Pb)_M = ({}^{206}Pb/{}^{204}Pb)_A\,(Pb_A/Pb_M)\,(Pb_M - Pb_B)/\,(Pb_A - Pb_B) + ({}^{206}Pb/{}^{204}Pb)_B\,(Pb_B/Pb_M)\,(1 - (Pb_M - Pb_B)/\,(Pb_A - Pb_B)) \quad (6)$$

which rearranges to:

$$({}^{206}Pb/{}^{204}Pb)_M = Pb_A\,Pb_B\,[({}^{206}Pb/{}^{204}Pb)_A - ({}^{206}Pb/{}^{204}Pb)_B]/Pb_M(Pb_A - Pb_B) + [Pb_A\,({}^{206}Pb/{}^{204}Pb)_A - Pb_B\,({}^{206}Pb/{}^{204}Pb)_B]/\,(Pb_A - Pb_B) \quad (7)$$

This is an equation of the form:

$$({}^{206}Pb/{}^{204}Pb)_M = k\,(1/Pb_M) + c \quad (8)$$

which is a linear equation between $({}^{206}Pb/{}^{204}Pb)_M$ and $(1/Pb_M)$, where the gradient is given by:

$$k = Pb_A\,Pb_B\,[({}^{206}Pb/{}^{204}Pb)_A - ({}^{206}Pb/{}^{204}Pb)_B]/(Pb_A - Pb_B) \quad (9)$$

and the intercept:

$$c = [Pb_A\,({}^{206}Pb/{}^{204}Pb)_A - Pb_B\,({}^{206}Pb/{}^{204}Pb)_B]/\,(Pb_A - Pb_B) \quad (10)$$

To summarize, the lead isotopic ratio of the mixture is a linear function of the reciprocal of lead concentration in the mixture. Thus, mixing lines in a two component mixture appear as straight lines when $({}^{206}Pb/{}^{204}Pb)$ is plotted against $(1/Pb_M)$. Similar equations apply for the equivalent relationships for ${}^{207}Pb/{}^{204}Pb$ and ${}^{208}Pb/{}^{204}Pb$.

Chapter 7

The FLAME GIS-Database

One of the aims of the FLAME project is to collect as many chemical and isotopic analyses of archaeological metal objects as possible from the whole of Eurasia, from approximately the fifth millennium to the first millennium BCE, but mainly focussing on the third and second millennia. When completed, it will be the largest publically-available database of archaeological Bronze Age metal analyses. This requires a database structure that is both flexible enough to collect a great deal of heterogeneous data, but simple enough for non-database experts to use. The structure of the database reflects the nature of the available information. The database is also connected to a GIS system, with an online interface that will be available at the project webpage https://metals.arch.ox.ac.uk. The interface between the database and GIS provides a simple and clear visualisation of the data available, according to their location, date, typology and archaeological context, as well as providing interactive tools which allow for queries, geostatistical analysis, and the downloading of data and results.

The database: general concepts

This is a description of the core structure of the database. The visualization of data in the database will vary according to the research questions, but how the information is acquired and stored will remain fixed in the structure described below.

The database has been created using PostgreSQL: a free, open-source object-relational database system. The structure has been created to be compatible with CIDOC-CRM, which is a standard (ISO 21127:2006) that "provides a semantic framework to describe the implicit and explicit concepts and relationships formalised into a database system" (http://www.cidoc-crm.org/). CIDOC-CRM has been largely used in cultural heritage documentation, in particular in museums and collections, but in recent years it has also proved its versatility in conceptualising scientific database structures. Building up a database according to a recognised standard has been a core principal in this project. Firstly, it provides a "test" of the quality of the database structure. Without this, many conceptual mistakes can become embedded into the mind of the creator of the database—for example, a single field in the database might not adequately reflect the complexity of the detailed information that is sometimes available. The use of CIDOC-CRM also has another important consequence for the research: it provides a guarantee of the preservation and usability in the future. All the database structures can be encoded, exported and transmitted in XML language. But the decision taken to structure a database in this way may not be clear *within* the structure (for example, what we mean by a particular field definition such as "site" or "context"). CIDOC allows us to encode within an XML code all the implicit reasoning that is displayed in the database structure. So the structure can be transmitted and re-used by other researchers in an unambiguous way, assuring its preservation into the future.

Some cardinal rules have guided the decisions taken when the structure was built:

1) The database is built with two different purposes:
   a) The database is a storage medium. It has the aim of including as much information as possible about metal artefacts from the whole of Eurasia. This information, spread across a number of different publications, and written in several languages, is not always easy to find. It will be united into one organic structure available to the public through the online portal. At first, everything will be offered in English but, ultimately, we are planning to provide Chinese and Russian translations.
   b) The database will support analysis. It will include tools that researchers can use not only to retrieve data, but also to undertake their own analysis of the underlying data using the methods developed for FLAME and described in the previous chapters.

2) The structure of the database is hierarchical. At the core of our database is a single metal artefact. Information is recorded not just for the artefact, but for the context, site and region from which it was derived and for samples taken from and analyses performed on the artefacts (see **Figure 1**).

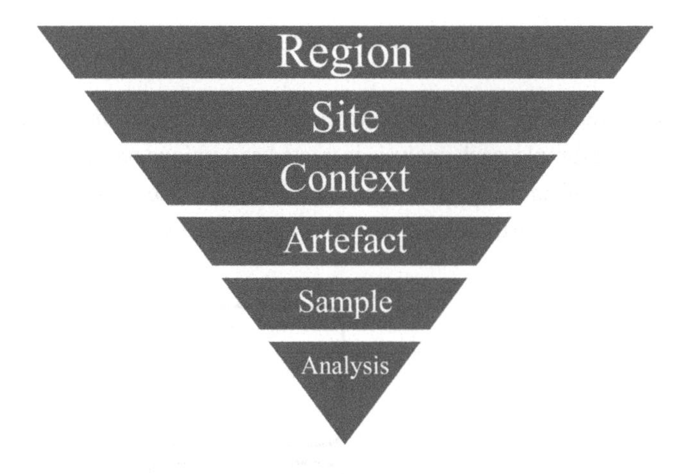

Figure 1:
The scales at which the database works.

3) The database is flexible: it is able to record very heterogeneous information, both in terms of its nature (archaeological, historical, artistic and scientific), and its quality. Most importantly, information that might be considered of "bad quality" is not excluded *a priori*. The original information is graded according to its reliability. Different factors may influence the reliability, including the analytical technique, with pXRF being rated the least reliable, as explained in Chapter 3. Another factor can be the opinion of the author of the publication from which the data are taken, as stated in the publication, or given via personal communication. We can also downgrade the reliability of analysis whose analytical total is not within the range of 95-105% (if the total is meaningful: see Chapter 3). Internal comparisons within the database can also highlight flaws in a specific set of analyses, as illustrated in Chapter 3 by the discussion of the analyses of nickel provided by von Bibra (1869).

4) The database is secure. There are different levels of authority that grant the permissions necessary to consult and use the database. This is particularly important because of the nature of the data, some of which may be considered to contain sensitive information (such as, for example, the precise location of a cemetery that is still under excavation). A system of "traffic lights" has been set to control the circulation of data. At the moment the data are available only to the group partners, but we are ultimately aiming to make it accessible to the public for consultation, keeping the possibility to modify and delete data only within the Flame research group.

The database: a description of the structure

The database is relational, but it has borrowed some concepts from object-oriented programming—example, each analysis is treated as a unique object independently of its nature (chemical analysis, isotopes, radiocarbon date, …). It is composed of over sixty tables, mostly related to each other by many-to-many relationships (**Figure 2**).

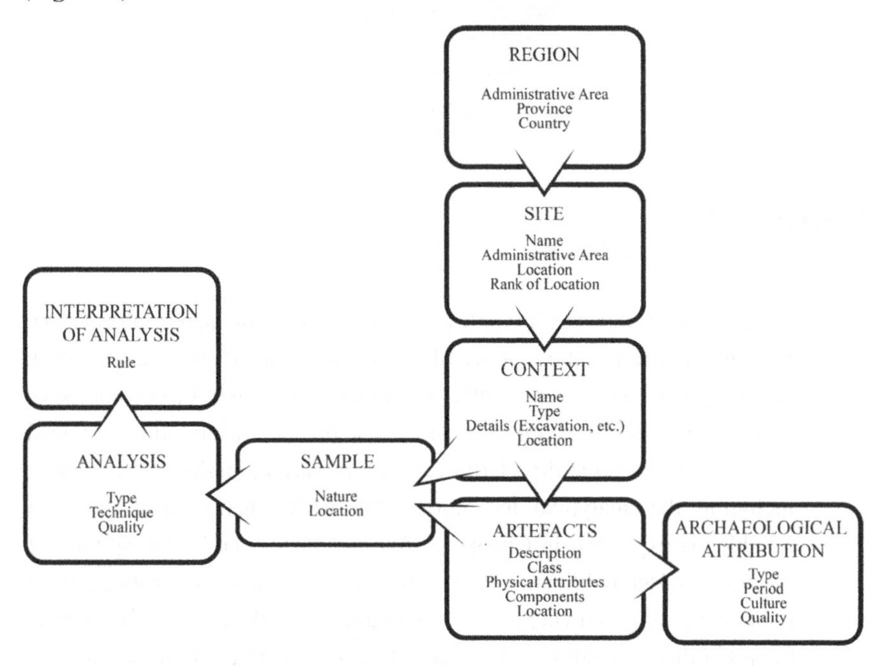

Figure 2:
A schematic diagram of the database structure.

The highest hierarchical level of the database is the *site*, where the key information resides. The *name of the site* is provided as a one-to-many relationship, so that is possible to assign different names to the same site. This is necessary for two main reasons: firstly, the same site may have been referred to in different ways in the literature. Secondly, as we are operating on a continental scale, with many different languages and alphabets, we are allowing the possibility of having the same site to be written both in its original language (and alphabet) as well as in the Roman alphabet.

The *site* table (**Table 1**) has a field dedicated to the *administrative area*, defined as the smallest modern political unit where the site is located. In some cases, the specific site of recovery of the material is not known. In these cases, usually a generic regional indication of provenance is provided. In such cases, this generic location is repeated in this field. The *administrative area* is recorded as it is known in English. The field "*site details*" is available for the additional details about the site that cannot be recorded in any other field. Since the coordinates may be a sensitive piece of information, in particular for those sites currently still under excavation, they will not be publically available for all sites. We also provide a *rank of location*, which indicates if the available coordinates refer precisely to the site, or only to the closest administrative area, or, more generically, to the region. This field is particularly important to assure the multi-scalar attribute of the database. Thus, if the user decides to operate on the level of the whole of Eurasia, all the analyses will be included, but when a study is dedicated to a very specific region—a province in France, for example—only those sites whose location has a scale of uncertainty smaller than the study area are selected.

Field	Type of field
Site ID	*text*
Site name	*text*
Parish	*text*
Province	*text*
Region	*text*
Country	*text*
Rank of location	*domain (text)*
Coordinates	*numeric*
Site details	*text*

Table 1:
List of fields in the table of sites.

The next level is given by the *context* (**Table 2**). The distinction between site and context is vital in order to operate with radiocarbon dates, since a clear record of the sequence of layers within the same site is required in order to undertake Bayesian statistical analysis. The *name of the context* and *context details* provide the equivalent information for the context as do site details and site name for sites in the *table of sites*. For the *context*, there is also the possibility to add information about the excavation (who excavated it, when, etc.…) and, in the case of a cemetery, the number of tombs, and any relevant information about the person buried, such as age and sex. A context is also defined by its *type*, which ranges within a fixed number of possibilities. Some are generic (burial, settlement, hoard, single find), and others are more specific (finding in water, lake dwelling, high altitude settlement, cave settlement, cave burial, ritual place, wall). When the quality of the information allows, it is possible to record the specific coordinates of the single context, in addition to the ones for the site. In this way, it becomes possible to undertake spatial analyses within a single site: for example, the chronological relationship between single graves of a cemetery can be highlighted, as in the cases of Remedello (de Marinis 1997) or Geimenlabern (Neugebauer 1991).

Field	Type of field
Context ID	*text*
Type of context	*domain (text)*
Date of Excavation	*text*
Institution that has excavated	*text*
Coordinates	*numeric*
Context details	*text*

Table 2:
List of fields in the table of contexts.

The *context* table has a one-to-many relationship with the table of *artefacts*: in one context there might be more than one artefact, but one artefact is related only to one context. The metal artefact is the core of the entire database, around which everything else is built. Defining what is meant by a *metal artefact* is not always intuitive, in particular in the case of composite artefacts. Is a necklace a unique artefact or are its single beads artefacts in themselves? The simplest and most effective solution is to record both the composite artefact and its parts as unique artefacts. There is a specific 'table of components' that allows the link to be made between the composite artefact and its components. In turn, within the table of artefacts, a component can be indicated as part of another artefact in a specific field. In this way, the fact that the analyses undertaken on the single components belonging to the same artefact can easily be ascertained.

Some information linked to the artefacts are straightforward, in particular the metric information (dimensions, weight), the presence of inscriptions or decoration, their current location in Museums or Collections, and any illustrations or pictures provided. Other information is more complex and is provided through a series of tables linked to the table of artefacts. It can be divided into archaeological and scientific information.

Archaeological information

This information typically relates to the typology, the chronology (mostly typo-chronology) and the affinity of the object to an archaeological culture.

Chronology

Chronology—the date of the artefact or assemblage—is critical to a project which aims to cover all of Eurasia, but it poses many difficult issues. Each region tends to have its own chronological system that varies greatly in terms of refinement, dating techniques used and the history of research. This is not only because of the vastness of the area of research, but also because the data we are compiling may have come from work done more than 50 years ago, using chronologies which have since been superseded. Most researchers, both European and Asian, tend to use locally-defined labels to indicate a time period, such as "Early Bronze Age" or "Chalcolithic", often sub-divided in an intricate and increasingly localised way. Defining what each author actually means by the labels that they use in terms of chronology is not an easy task. In fact, it is an enormous task, which has been the focus of a project that has been running for many years, the Period-O project (http://perio.do/), whose concepts are taken as guidelines for the present database.

Firstly, the creation of a *table of periods* is required. This includes all the labels used by the different authors, and assigns to each label a time span: a *post quem* and an *ante quem*. This assignment is done by reference to a published chronology, such as, for example, the one published by the Römisch-Germanisches Zentralmusem, available online at http://odur.let.rug.nl/arge/Work/chrono.htm. It is important to understand that the same label, such as "Bronze Age", has a completely different chronological meaning according to location. "Bronze Age" in the British Isles has a different time span from "Bronze Age" in Turkey (respectively 2200–800 and 3300–1200 BCE). For this reason, the geographical location is also recorded when entering a period in the database.

The connection between the table of artefacts and the periods is a many-to-many type. In fact, many artefacts are dated to the same period, but it is also true that the same artefact may have a different chronological definition, according to different authors. The link between the artefacts and the periods occurs through an intermediate table called "chronological attribution" that indicates which author determined that an object belongs to a particular period, in which publication this attribution is presented, and how reliable this attribution is. The reliability is initially set as "average" for all the assignments, but, if internal inconsistencies emerge (for example, if the same author gives different attributions to objects of the same type), the reliability will decrease. The maximum grade of reliability of chronological assignment is assigned to those derived from radiocarbon analyses undertaken on samples taken directly from the artefact itself (such as from a bone or wooden handle still attached to the metal object). The second highest grade attribution links objects and radiocarbon dates of the context from which the objects are found. An internal script calculates the maximum time-span into which the object falls, taking into consideration all the chronological attributions and their reliability. Firstly, it picks up the reliabilities that have the highest score. More than one reliability may have the same high score. Then, among the dates selected, it calculates the minimum *ante quem* and the maximum *post quem*. For example: the same object may have three chronological attributions of 3300–2500 BCE, 3000–2200 BCE and 3500–2800 BCE. The first two have been graded 7, and the last one 5. The time span in which the object falls is then determined to be 3300–2200 B.C. This calculated time span is recorded as an attribute in the table of artefacts, in the fields *post quem* and *ante quem*.

The use of a time span to indicate of the date of an artefact has the advantage of being universally recognised. *Post quem* and *ante quem* have the same chronological meaning in Portugal as in the Altai region. But, of course, the user needs to interrogate the database in a meaningful way. The resolution of the chronology is dependent on the geographical region of Eurasia, and this needs to be taken into consideration. This means that the database can be used to understand the evolution of a specific phenomenon or process, such as the adoption of alloying techniques in contiguous regions, but the time slices used to undertake the analysis need to greater than the precision of the time assignment in that region. So, if in Turkey the chronological system has only a precision of 500 years, whereas in Greece there could be a precision of 50 years, the analysis of the evolution on the use of bronze in both Greece and Turkey cannot use time slices smaller than 500 years.

Typology

The assignment of an object to a typology poses similar problems as those discussed above for the chronology. A field exists that provides a description of the object and also a field which has its translation into English, if available. But a table of types has also been created. This includes the object types recognised in the typologies proposed by different authors. Each metal type is possibly linked to many objects. Similarly, the same object may be linked to different types, according to the interpretation of different authors. So, another table of attribution, in this case of types, has been created. As with the one for chronologies, this table records who made the attribution, the publication where this attribution took place, and a reliability score of the attribution, which, as a start, is set as 'average'. The same structure has also been created to record the cultural attribution of the artefacts.

With the objective of facilitating querying the database, in addition to the description of the objects and the typological attribution, a more generic "class of object" has been provided as a field for each artefact, which is broadly related to the function of the object. This attribute has a controlled domain made of fixed words. The creation of this domain has required the integration of different typologies and different approaches to the material culture in different periods and locations. These difficulties need to be taken into consideration by the user of the database. So, for example, the term 'vessel' is used to describe something completely different in the European Bronze Age and in the Chinese Shang and Zhou periods. This difference is in not only manifested in their shape and function, but also affects their physical dimension and weight, their decoration, their production techniques (usually sheet hammering in Europe, but casting in China), and, above all, the conceptual meaning of the object in both the producing culture and the consumption culture (which may or may not be the same). Some of this information is recorded in the database (in particular the physical data, where available), but it is necessary for the user to understand that searching for "vessel" across the whole of Eurasia will give a number of objects that may have in common only a very generically attributed shape or function.

Scientific analyses

Sample

A scientific analysis is strictly undertaken not on an "artefact" but on a *sample* of that artefact, even if a sample is not physically removed. Each artefact may have more than one sample taken, either because it may consist of several different parts, or simply because it is multiply sampled, and this can result in several analyses of the same object. When available, information about the sample is recorded: its *position*, *dimensions*, *state of preservation*, the *legal owner,* and its *current whereabouts*. This information is very rarely to be found in publications but is a factor to be taken into consideration when evaluating the representativity of the sample and, ultimately, the quality of the analysis. The issue of the chemical inhomogeneity of copper alloy objects is discussed in Chapter 3, along with some recommendations for best practice when analysing and publishing such objects, or 'analytical hygiene'.

The database also aims to record, whenever possible, information about radiocarbon dates associated with the metal artefacts. Conceptually, a sample is a sample, independently of the analysis that can be undertaken on it, so there is only one table of samples both for chemical and isotopic analyses, and also for radiocarbon analysis. The sample for the radiocarbon dating can be taken from organic residues on the metal object or from the context associated with the artefact, therefore the table of samples is linked not only to the table of artefacts but also to the table of contexts. A specific field in the table indicates the *nature of the sample*: i.e., if it is metallic or organic, and, if organic, whether it is taken from bone, tooth, wood, charcoal or carbonised remains.

Analysis

The table of analytical results stores the information about the analysis taken from each sample, independently of the nature of the sample (i.e., chemical and/or isotopic analysis of metal, or radiocarbon dating of organic material). Some fields are common to all types of analyses: the *laboratory* where the analysis has been undertaken, *laboratory number*, *date of the analysis*, *run quality*, *technique used*, *legal owner* and *bibliography*. Some other fields are specific to the nature of the analysis. The typical information from lead isotope analysis are at least three sets of the ratios of ^{204}Pb, ^{206}Pb, ^{207}Pb and ^{208}Pb (preferably as given by the instrument, rather than re-calculated—see Chapter 6). Radiocarbon analyses are recorded with their uncalibrated value, the standard deviation of the measurement, and the d^{13} value of the sample. Metallographic images or descriptions are possibly the hardest type of analysis to record into a database, because of the descriptive nature of the results. In addition to the free-text "description" field, some quantitative information is recorded, such as the dimension of the grains, the presence of dendrites or of twinnings, the hardness value and the technique used to calculate these parameters.

The information of the chemical analysis is recorded into the database **exactly as provided in the original dataset**. This means that we aim to have an exact copy of the original publication, including any semi-quantitative symbols used (such as 'tr', '-', 'nd' or '<0.05'), and any errors. It is important to recognise that the nature of the original data is a carrier of valuable information *per se*: it can reveal a great deal about the history of research, the techniques used to analyse archaeological metals, and about what in the past has been considered to be a "good" analysis.

Interpretation

As mentioned, the database is not only a vehicle for the storage of data, but is also a functional, active instrument for research. This means that the database is “doing” something to the original data, and this action, too, needs to be properly recorded. The first basic operation that the database does is to transform the above-mentioned uncertainties in the original data into numeric values, which can be further treated using statistical and geo-statistical analysis. Chapter 3 explains how we deal with these issues when we attempt an analysis of the data. The result of this modification is recorded in the table “*interpretation of chemical results*”, where the methodology used to extract the numerical data is also recorded.

Radiocarbon dates can also be processed, in particular for the calibration of the dates. So we provide a table that records the calibrated date, the curve used for the calibration, and the software used (typically the latest available version of OxCal: (https://c14.arch.ox.ac.uk/oxcal.html)).

Statistical and geostatistical tools

The database has been published as an online GIS service, together with custom tools created to interrogate and analyse the data. On the web-site both the analytical data as published and the numeric chemical data are available. The first and simplest kind of analysis that can be undertaken is the querying of the data according to the chronological span, type of object and type of site. The result of the query is visible as an online table and can be exported as a csv file. The considerations discussed above about the assignment of objects to period and typology should serve as a warning. The database is a useful tool for research but is not intended as (and could never be) a black box that can give all the answers to archaeometallurgical questions without a proper interpretation by the user. In the example of the definition of ‘vessel’ outlined above, understanding the conceptual difference amongst an assemblage of artefacts selected according to their broad category is vital. We provide data summary routines such as histograms, pie charts, cumulative frequency curves and kernel density estimates. Some other tools, specifically created for this project, are offered.

Normalise

The *Normalise* function for chemical data acts in two stages. The first step is to calculate the percentage of copper in those cases where it was not reported in the original analysis (such as in the SAM database). This is simply done by summing all the weight percentages of the elements that have been analysed and subtracting the result from 100. The second step is to subtract those elements that are considered to be intentionally added so that the trace elements associated with the copper can be re-calculated (see Chapter 4). Deliberate addition of alloying elements will have a dilution effect on the percentage of the trace elements in the copper used to assign an object to a Copper Group (As, Sb, Ni and Ag: see Chapter 4). Tin, zinc and lead are removed from the analysis, since they are considered to be 'intentionally alloyed' (even if present at <1%) rather than associated with the copper source. Arguably if they are present at less than 1% they could be considered to be part of the trace element suite associated with the copper ore, and thus left in the analysis, but in order to obtain a systematic classification of the copper, all the (potentially) alloying elements are removed from the composition. See also Chapter 4 for a discussion of the role of As as an alloying element. The results are shown in an operational layer whose attribute table has a series of fields, one for each element, with the resulting value normalised, according to the formula:

$$Xnormalised = x * \frac{100}{TOT-(Sn+Zn+Pb)}$$

Grouping

The *Grouping* tool takes as input the resulting operational layer of *Normalise,* and allocates each artefact to a Copper Group (see Chapter 4) based on the presence/absence of As, Sb, Ag and Ni. The pre-set threshold values of the tool are the "typical" ones of the FLAME method, namely 0.1%, but the threshold to define the presence of an element can also be set by the user of the program, by filling in a dedicated field.

Parameters of the tool	Nature of the parameter	Predefined value
Input File	Table of the analysis with numeric values	-
Limit As	Threshold to evaluate the presence of As	0.1
Limit Sb	Threshold to evaluate the presence of Sb	0.1
Limit Ag	Threshold to evaluate the presence of Ag	0.1
Limit Ni	Threshold to evaluate the presence of Ni	0.1
Output File	Resulting table where a field with copper group has been added	-

Table 3:
Parameters required from the user for the Grouping tool.

The result is an operational layer with a field that shows the Copper Group to which each object belongs. If the user is interested in the simple distribution maps of the resulting Copper Groups, they can use the tool "*Create Maps of Groups*." This tool has as input the resulting operative layer of the "Grouping tool" and creates a series of sixteen distribution maps, one for each Copper Group. But, as the FLAME method has demonstrated in several case studies, the most useful information may be obtained using *Ubiquity Analysis*, namely calculating the percentage of objects with a specific composition in a determined area.

Ubiquity analysis

This analysis calculates the percentage of the frequency of a specific composition over the total of the assemblage, as outlined in Chapter 4. The use of GIS can introduce a spatial dimension into the definition of a metal assemblage. If we imagine space as divided into a sequence of polygons, or a net, we can define a series of metal assemblages as being all the metal artefacts that have been found in each of the polygons of the net. Mapping the Ubiquity Analysis within each polygon of the net gives a nuance of the variance of "importance" of one composition in space. Ubiquity analysis can be done by taking into consideration one compositional group over all the possible Copper Groups resulting from the Grouping tool, but it can also evaluate the percentage of objects that have an element (e.g., Sn) above a certain percentage (e.g., 1%). The tool *Ubiquity of Copper Group* (see **Table 4**) is designed for the first case, and *Ubiquity of Element* (**Table 5**) for the second.

Parameters of the tool	Nature of the parameter	Predefined value
Input File	Table with a field dedicated to the compositional groups	-
Aggregator Net	Shapefile of a geographic net. Each single polygon of the net delimitates a metal assemblage	-
Output File	Shapefile with graduated colours that reflect the different frequency of a compositional group for each polygon of the net	-

Table 4:
Parameters required from the user for the Ubiquity of Copper tool.

The Ubiquity of Element tool allows us to evaluate the presence of alloying elements in metal—a useful piece of information to evaluate the perception and the use of metal in ancient societies. It gives the user the possibility to choose one specific element for which they want to analyse the ubiquity (e.g., tin), and also to set both the lower and upper threshold of presence in which they are interested (T**able 5**). The tool is pre-set to analyse the presence of an element with a percentage between 1% and 100%.

Parameters of the tool	Nature of the parameter	Predefined value
Input File	Table of the analysis with numeric values	-
Element	Element whose presence is evaluated	-
Upper Limit	Lower threshold of presence of the element	1
Lower Limit	Upper threshold of presence of the element	100
Aggregator Net	Shapefile of a geographic net. Each single polygon of the net delimitates a metal assemblage	-
Output File	Shapefile with graduated colours that reflect the different frequency of objects with the presence of a specific element for each polygon of the net	Required

Table 5:
Dialogue box for the Ubiquity of Elements.

Both the Ubiquity of Copper Groups and Ubiquity of Elements tools require the definition of the unit area for which the ubiquity is calculated. Setting the spatial bin of the analysis has important consequences, and is known in the literature as MAUP (Modifiable Areal Unit Problem: see Harris (2007)). There is no perfect solution for this problem, and the user is provided with different options. The strongest results are the ones that are coherent even with the use of different Unit Areas. At the moment, the available unit choices are:

- *single sites*: with this option, the percentage of presence of each compositional group is calculated site by site. The use of this option is not recommended when the study area has too many single finds, because the result is composed of many hotspots where one chemical composition represents 100% of the whole of the assemblage.
- *river basins*: this option may highlight the importance of the river system in the circulation of metal. The disadvantage is that the areas of river basins are highly variable, and so the study areas can be highly variable.
- *one degree on the WGS84 coordinate system*. With this method, the consistency of the unit area is assured. On the other hand, in those regions where the distribution of the metal artefacts is sparse, there may be several squares that are not populated.

The use of GIS is not limited to the production of distribution maps, but allows us to insert topographic variables into the analysis of the distribution of metals. When we talk about metal flows and movement of metal we are, ultimately, talking about the movement of people, which occurs in a three dimensional world where topography, river systems, and vegetation may be determinant factors. For example, in Perucchetti *et al.* (2015) we demonstrated the importance of the river systems in the movement of metal in the prehistoric alpine region. River basins and find spots of metal artefacts (this information being available) have been mapped together and used to verify the correlation between river basins and compositional groups. The χ^2 test suggested that the distribution of different metal groups was not random with respect to river basins, neither in the Copper Age nor in the Bronze Age A1, with a p-value of less than 0.005 in both cases (**Figure 3**).

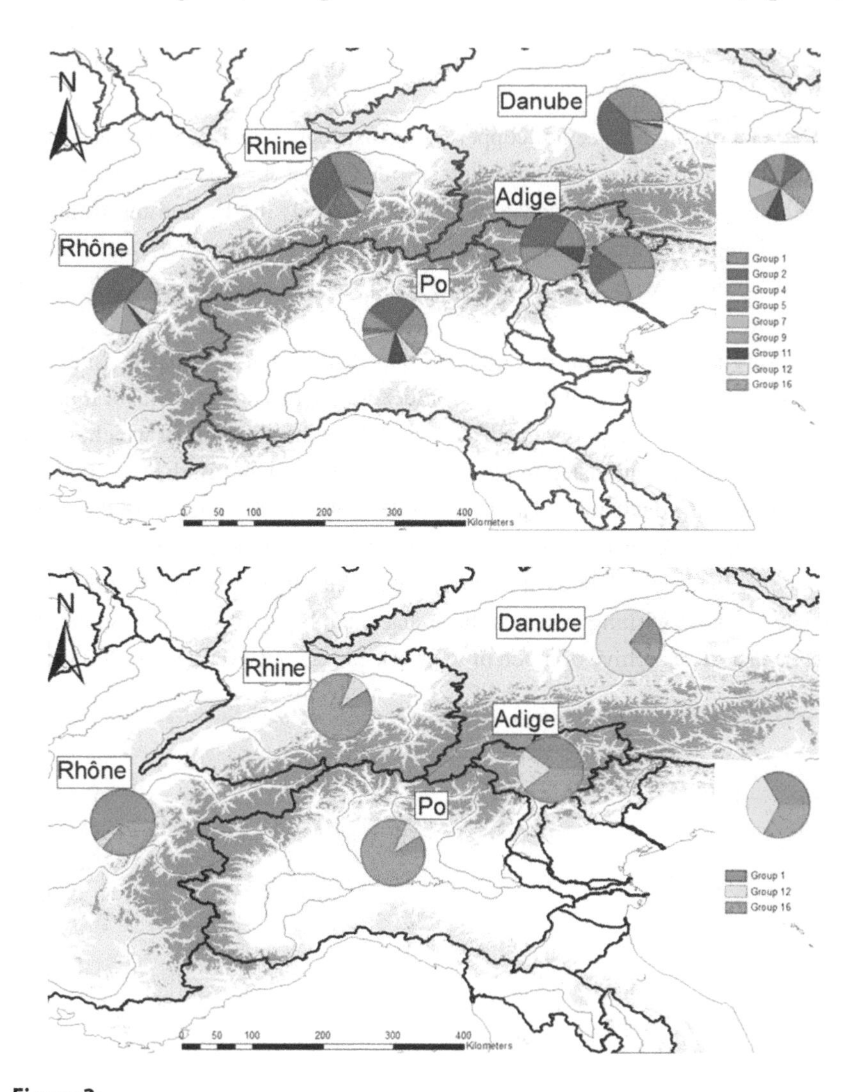

Figure 3:
maps showing the presence of different Copper Groups (as defined in Chapter 4) in different river basins in the Circumalpine Region. Above: how the Copper Groups were distributed in the Copper Age (3600–2200 BCE); below in the Early Bronze Age 1a (2200–2000 BCE) (Perucchetti 2017).

Figure 3 shows maps of the river basins in the Alpine region. The pie charts represent the distribution of different Copper Groups, as defined in Chapter 4, in the Copper Age (3600–2200 BCE: upper figure) and Early Bronze Age 1a (2200–2000 BCE: lower figure). In the Copper Age a larger number of Copper Groups were present, but their distribution was not even with respect to the river basins. For example, the Rhone and Po river basins have a predominance of Copper Group 2 (copper with arsenic>0.1%), whereas the Rhine and Danube have a broadly equal presence of Copper Group 1 and 2 (respectively copper without impurities and copper with arsenic).

The relationship between of copper group distribution and river basins is even clearer in the subsequent period, where fewer Copper Groups are present. The Rhone valley has a predominance of Copper Group 1; the Po and the Rhine of Copper Group 16 (copper with arsenic, antimony, silver and nickel), and Copper Group 12 (copper with arsenic, antimony and silver) dominates in the Danube river basin.

The varying distributions of Copper Groups in different river basins, suggests the use of rivers as a mode of transport for metal (either as raw material or finished objects). This is also supported by discoveries of longboats, some of which, found in the Alpine region, are dated to the Bronze Age (Ravasi and Barbaglio 2008).

Ultimately, the database and associated GIS tools will become one of the major outputs of the FLAME project. We are constantly revising and improving it, and continue to load new data as it becomes available. It will serve not only as a resource for those wishing to search the archaeometallurgy of specific regions, but also as a historical archive of much of the work carried out over the last 200 years.

Chapter 8

Summary: Beyond Provenance?

The title of this volume is intended to provoke thought—for archaeologists, to encourage thinking beyond individual objects to assemblages and populations, and, for archaeometallurgists, to think beyond the 'holy grail' of provenance (Radivojević *et al.* 2018). The concept of provenance has had a remarkable stranglehold on the discipline since its first conception in the 1860s. It appears to be definitive and 'scientific'—object A comes from metal mined at source B, on the basis of these trace elements, or this isotopic pattern. Everybody recognizes that it is exclusive rather than positive—it can only exclude potential sources, rather than make a positive attribution—and yet it continues to be sold to the archaeological and general public as a key concept. Possibly it has provided a convenient alibi for the involvement of aspects of materials science in archaeology, when in fact none was needed—perhaps by providing the answer to a question not asked, rather than simply viewing materials science as one aspect of the multidisciplinary activity that is archaeology, whose collective aim is the better understanding of the human past.

Where it is possible, of course, provenance provides a useful piece of information—a piece in the jigsaw puzzle—but what does provenance actually mean? In both metal and glass studies, it has long been realized that we need to distinguish between 'primary' and secondary' manufacture—primary being the actual manufacturing source of the raw material, and secondary being the place where this raw material is manipulated and converted into objects. In Roman glass studies, for example, the places of 'primary' manufacture, where sand and alkali are converted into raw glass, appear to be very limited indeed, whereas a large number of glass object production furnaces are widely spread across the empire (Degryse 2014). Interestingly, a high proportion of the glass analysed in their study (25%) appeared to have been recycled.

The concept of provenance indeed becomes 'fuzzy' when recycling and secondary production sites are considered. But is provenance anyway an entirely modern construct, which, because we can sometimes determine it, we believe to be important, and therefore project this importance back into the past? What would 'provenance' have meant to people in the past? We drive cars with components that come from all over the world, but we do not value them for that property. Perhaps metal, in the early stages at least, had a very strong symbolic property, representing power, magic, or some long distance contact with powerful people, which we might conflate with the concept of provenance. There is no doubt that if we can 'prove' that object A comes from metal mined at source B, then, as stated by Damour (1865, 313): "*it is inferred that there has been transport of the object itself, or at least of the matter of which it is formed*." The means by which this happened is of course the stuff of modern scholarship, although often it is indicated simply by arrows on a map—bereft of geographical hindrances, but, more significantly, happening in the complete absence of people. But did, for example, Early Bronze Age people in Britain even know that some or all of the metal in their axes might have originated in Ireland or Spain? Probably not. More likely, they were aware that it came from a group of people just over the horizon, and not that it represented the last link in a series of interactions which stretched over many hundreds of kilometres, and could have involved exchanges that took place over many decades. Did it matter to them is perhaps a more interesting question. In more centralized societies, complex supply networks would have supplied raw metal to foundries. The quality of the metal would have, in many cases at least, mattered, but was that a reflection of the provenance of the metal (an early 'branding' phenomenon—'metal from A is better than metal from B' for reasons of prestige, reputation, kinship, magic, etc.) or was it practical, based on working properties, colour, smell, etc. We have the privilege of, hopefully, being able to take an overview of all of these complex networks of interaction, but we must remember that what we see is not what they saw, and be careful when attributing knowledge and motives to people in the past.

In this volume we have argued that provenance, as traditionally interpreted, is an important component of the broader picture of the biography of objects, and the prosopography of assemblages, but certainly not the only component, and perhaps not the most important. What happens to the metal after it has been mined and smelted is key to understanding the interaction between humans and metal. It is likely to be alloyed (perhaps at a secondary site), traded or exchanged in some way, and in some cases passed on in one form or another. It might then be melted down and re-used—one person's 'axe' is another person's 'ingot'—if the original axe has no further symbolic or cultural meaning, or if, conversely, there is symbolic value in incorporating that metal into a new object. We have introduced here the concept of the 'flow' of metal—the underlying parent population, from which all archaeologically recovered objects are abstracted. This is essentially a theoretical construct, the composition of which is inferred from the analysis of a sub-set of the archaeologically recovered objects, that allows us to link conceptually the 'inputs' into the system (mining sources, known and unknown, plus recycled material) with the 'outputs', in the form of archaeologically recovered and chemically analysed objects. In all cases, the unit of study is the assemblage rather than individual objects.

We have derived a whole new set of tools to carry out this study. This is for two reasons, one related to the nature of the data, and one to the aims of the study. Our aim is to use patterns of metal circulation to infer the multiple sets of relationships which existed across Bronze Age Eurasia. To do this we need the largest possible database, and that means assembling chemical data derived over the last hundred years, and possibly even earlier. Consequently, we have a very heterogeneous database, with analyses done by different analytical methods and by different research groups, and therefore with different levels of accuracy and precision. Quite clearly, the application of sophisticated methods of data analysis involving clustering and discriminant analysis (chemometrics) to such a heterogeneous database is more likely to cluster by analytical method or analyst, rather than by something archaeologically meaningful. Rather than abandon all this earlier information, we have chosen to adapt our interpretative methodology to deal with imperfect data.

In terms of the aims of the study, and in particular to look beyond the idea of provenancing metal, we have effectively gone back to 'archaeological basics'. Archaeology is fundamentally about deducing human behaviour from changes observable in the archaeological record. In our case, the archaeological record is the corpus of analysed metal objects from the Eurasian Bronze Age, and the tools described here are designed to initially highlight changes in this record. Such changes might be a consequence of differences in human behaviour and/or the availability of resources over time, as one period succeeds another, or it could be the result of different practices at different places, revealed by comparing the metalwork of contemporary societies. There are always complexities within such comparisons which need to be considered, such as that due to the metal used in weapons and vessels being different within a particular society, or that the metal deposited as hoards might be different (typologically and chemically) from the metal used for objects deposited in tombs. Nevertheless, the first step is to detect change (or, indeed, continuity), and then to hypothesise about what might have caused such change. It could be that the simplest explanation is a change in ore source, in which case we are back to the 'holy grail' of provenance, but in many cases it will be a conjunction of more complex phenomena.

We have taken a set of simple procedures as the starting point for these comparisons. The idea of Copper Groups (chapter 4) is that by using a presence/absence system to allocate objects to one of a set of 16 groups, and then aggregating all objects within an assemblage to form a ubiquity distribution across the 16 groups, we can make simple comparisons across several assemblages. These comparisons might be intended to reveal changes over time, if succeeding phases or periods are compared, or between contemporary sites. We never assume that a particular Copper Group is associated with a single source, or that individual sources can contribute to only one Copper Group. Although we normally use a cut-off value of 0.1% for the four trace elements (arsenic, antimony, silver and nickel), we recognize that this is a compromise designed to accommodate heterogeneous data, and do recommend variation according to the quality of the data.

Likewise, we eschew the classification of Bronze Age alloys into modern alloy categories (Chapter 5). This makes the implicit assumption that ancient metalworkers were striving to achieve alloys with modern characteristics of colour and working properties. In some cases, undoubtedly, they were aiming for some optimal product, which may or may not have been allied with modern perceptions of metal, but the reality of the data is that many compositions, for whatever reason, do not fall into such simple categories. We have therefore adopted a presence/absence system for the alloying elements (tin, lead and, rarely for the Bronze Age, zinc), with a default value of 1%. This is designed to give equal weight to the large number of objects which do not correspond to modern alloy definitions, and which might contain much hitherto overlooked information about the biographies of these objects—recycling, re-alloying, etc.

Perhaps the most radical methodological contribution relates to the interpretation of lead isotopes (chapter 6). The conventional use of lead isotope data in archaeology is based on the method of presentation used in isotope geochronology—the plotting of a pair of isotope diagrams, based around some combination of three measured isotope ratios. In geology this is usually $^{207}Pb/^{204}Pb$ vs. $^{206}Pb/^{204}Pb$ and $^{208}Pb/^{204}Pb$ vs. $^{206}Pb/^{204}Pb$, but the practice varies in archaeology. Such diagrams were originally devised to allow the geological age of the deposit to be determined graphically, but, we would argue, are not necessarily suited to the display of data from archaeological objects. The additional complexity of the possibility of anthropogenic mixing of lead from different sources, or of adding lead to copper, renders conventional geological lead isotope ratio diagrams somewhat unsuitable for use in archaeology. We have proposed a new methodology based on that used in strontium isotope geology, which is to plot the inverse of the lead concentration in the object (1/Pb) against one of the three possible isotope ratios. Although now there are in principle three different figures to look at, we have found them extremely useful for detecting the mixing of leads with different isotope ratios, and also for identifying the addition of lead to copper, since the diagram contains compositional as well as isotopic data. We have also experimented with deriving the geochemical 'fundamental parameters' of a lead deposit – μ (the ratio of $^{238}U/^{204}Pb$), κ (the ratio of $^{232}Th/^{238}U$), and T (the 'model age' of the deposit). These are three independent parameters which together characterize the geochemistry of the deposit. Finally, we have shown how the old question of 'how do we isotopically define an orefield', which bedevilled much of the literature in the 1990s, can now easily be resolved using kernel density estimates. Moreover, using these probability distributions, we can test different assemblages for the degree of overlap between them, which allows rigorous comparisons to be made for the first time.

One interesting conclusion from this work on lead isotope data from archaeological material is that we believe that the major conclusions can be derived from relatively low resolution isotope ratio data—many of the plots reveal archaeologically meaningful structure by looking at only the first two decimal places in the measurements. That is not to say that high resolution measurements should not be made, or that under particular circumstances the full resolution of the data might not be necessary to reveal the fine detail within the data. It is simply an observation that for archaeological material, where there is the potential for large variations in the data due to mixing, and because of the inherent variability within the sources themselves, these low resolution interpretations might be sufficient to reveal the archaeological story.

Numerous scholars since Göbel (1842) have emphasized the potential to look beyond the chemical data, and beyond the metals themselves, to reveal patterning that is evidence for human behaviour. Thus the urge to put metal chemistry into a social context is not new. Perhaps the most influential of these scholars has been Chernykh, but many others, including Ottaway, have attempted to use chemical and isotopic data to say something about the people who made and used the metal, rather than answer questions such as provenance, which, in isolation, are quite abstract. The aims and objectives of the FLAME project are therefore not new, apart from, perhaps, the scale. What is new, however, is the attempt to practically achieve these objectives by deriving a new set of tools to deal with large scale data of variable quality, and to look at these data in a new way.

References

Albarède, F., Desaulty, A.M. and Blichert-Toft, J. (2012). A geological perspective on the use of Pb isotopes in archaeometry. *Archaeometry* **54**, 853–867.

Anderson, C.T. (1930). The heat capacities of arsenic, arsenic trioxide, and arsenic pentoxide at low temperatures. *Journal of the American Chemical Society* **52**, 2296–2300.

Anon. (1946). Early Mining and Metallurgy Group. *Man* **46**, 99.

Bagley, R.W. (1987). *Shang ritual bronzes in the Arthur M. Sackler Collections*. Arthur M. Sackler Foundation, Washington, D.C.

Barnes, I.L., Shields, W.R., Murphy, T.J., and Brill, R. H. (1974). Isotopic analysis of Laurion lead ores. In *Archaeological Chemistry*, ed. Beck, C.W., Advances in Chemistry Series 138, American Chemical Society, Washington, DC, pp. 1–10.

Baxter, M.J. and Gale, N.H. (1998). Testing for multivariate normality via univariate tests: a case study using lead isotope ratio data. *Journal of Applied Statistics* **25**, 671-683.

Baxter, M.J., Beardah, C.C. and Wright, R.V.S. (1997). Some archaeological applications of kernel density estimates. *Journal of Archaeological Science* **24**, 347-354.

Beeley, P. (2001). *Foundry Technology*. Butterworth-Heinemann, Oxford.

Bergman, T.O. (1777). Disquisitio chemica de terra gemmarum. *Nova Acta Regiae Societatis Scientiarum Upsaliensis* **3**, 137-170.

Bertholet, M. (1888). Sur le nom du bronze chez les alchimistes grecs. *Revue Archéologique* **12**, 294-298.

Bidegaray, A.-I. and Pollard, A.M. (2018). Tesserae recycling in the production of medieval blue window glass. *Archaeometry* **60**, 784-796.

Biek, L. (1957). The examination of some copper ores: A report of the Ancient Mining and Metallurgy Committee. *Man* **57**, 72-76.

Blades, N.W. (1995). *Copper alloys from English archaeological sites 400–1600 AD: an analytical study using ICP-AES*. Unpublished PhD dissertation, University of London.

Bradley, R. (1988). Hoarding, recycling and the consumption of prehistoric metalwork: Technological change in Western Europe. *World Archaeology* **20**, 249-260.

Bray, P.J. (2009). *Exploring the Social Basis of Technology: Re-analysing Regional Archaeometric Studies of the first Copper and Tin-Bronze use in Britain and Ireland*. Unpublished D.Phil. Thesis, University of Oxford.

Bray, P.J. and Pollard, A.M. (2012). A new interpretative approach to the chemistry of copper-alloy objects: source, recycling and technology. *Antiquity* **86**, 853-867.

Bray, P., Cuénod, A., Gosden, C., Hommel, P., Liu, R. and Pollard, A.M. (2015). Form and flow: the 'karmic cycle' of copper. *Journal of Archaeological Science* **56**, 202-209.

Brill, R.H. and Wampler, J.M. (1967). Isotope studies of ancient lead. *American Journal of Archaeology* **71**, 63-77.

Bronk Ramsey, C., Housley, R.A., Lane, C.S., Smith, C.V. and Pollard, A.M. (2015). The RESET tephra database and associated analytical tools. Quaternary Science Reviews 118, 33–47.

Brown, M.A. and Blin-Stoyle, A.E. (1959). A sample analysis of British Middle and Late Bronze Age material, using optical spectrometry. Proceedings of the Prehistoric Society 25, 188-208.

Budd, P. and Ottaway, B.S. (1990). Eneolithic arsenical copper: chance or choice? In *Ancient Mining and Metallurgy in Southeastern Europe*, ed. B. Jovanović, Archaeological Institute, Belgrade/Bor Museum of Mining and Metallurgy, pp. 95-102.

Budd, P., Haggerty, R., Pollard, A.M., Scaife, B. and Thomas, R.G. (1996). Rethinking the quest for provenance. *Antiquity* **70,** 168-174.

Caley, E.R. (1939). *The Composition of Ancient Greek Bronze Coins*. American Philosophical Society, Memoirs, Vol. 11, Philadelphia.

Caley, E.R. (1964a). *Orichalcum and Related Ancient Alloys*. Numismatic Notes and Monographs 151, American Numismatic Society, New York.

Caley, E.R. (1964b). *Analysis of Ancient Metals*. Pergamon Press, Oxford.

Caley, E.R. and Richards, J.F. (1956). *Theophrastus on Stones*. Ohio State University, Columbus, Ohio.

Carter, G.F., Caley, E.R., Carlson, J.H., Carriveau, G., Hughes, M.J., Rengan, K. and Segebade, C. (1983). Comparison of analyses of eight Roman orichalcum coin fragments by seven methods. *Archaeometry* **25**, 201-213.

Charles, J.A. (1980). Recycling effects on the composition of non-ferrous metals. *Philosophical Transactions of the Royal Society of London Series A* **295**, 57–68.

Chase, T. (1974). Comparative analysis of archaeological bronzes. In *Archaeological Chemistry*, ed. Beck, C.W., Advances in Chemistry Series 138, American Chemical Society, Washington, D.C.

Chen, K. (2009). 陕西汉中出土商代铜器的科学分析与制作技术研究 *Shaanxi Hanzhong chutu shangdai tongqi de kexue fenxi yu zhizuo jishu yanjiu* (Scientific Study on the Shang Dynasty Bronzes Unearthed from Hanzhong, Shaanxi Province: Materials and Manufacturing Techniques), University of Science and Technology Beijing.

Chernykh, E.N. (1966), Istoriia drevneĭsheĭ metallurgii Vostochnoĭ Evropy. *Materialy i issledovaniia po arkheologii SSSR* **132**, 1-145.

Chernykh, E.N. (1970). O Drevneyshikh Ochagakh Metalloobrabotki Yugo–Zapada SSSR. *Kratkie soobshcheniia Instituta Arkheologii* **123**, 23–31.

Chernykh, E.N. (1978). *Gornoe Delo I Metallurgiya v Drevneyshey Bolgarii*. Bulgarian Academy of Sciences Press, Sofia.

Chernykh, E.N. (1992). *Ancient Metallurgy in the USSR: The Early Metal Age*. Cambridge University Press, Cambridge.

Chernykh, E.N., Avilova, L.I. and Orlovskaya, L.B. (2000). *Metallurgicheskie provintsii i radiouglerodnaya khronolgiya*, IA RAN, Moscow.

Chernykh, E.N. (ed.), *Каргалы (Kargaly). Том I. Geological and geographical characteristics. History of discoveries, exploitation and investigations. Archaeological sites (2002). Том II. Gorny – the Late Bronze Age settlements. Topography, lithology, stratigraphy. Household, manufacturing and sacral structures. Relative and absolute chronology (2002). Том III. Archaeological materials. Mining and metallurgical technology. Archaeobiological studies (2004). Том IV. Kargaly' necropolis. Kargaly population: palaeoanthropological investigations* (2005). *Том V. Kargaly: phenomenon and paradoxes of development. Kargaly in the systems of Metallurgical Provinces. Hidden (sacral) life of archaic miners and metallurgists* (2007). Языки славянской культуры, Москва (2002-2007).

Chernykh, E.N. (2008). Formirovanie Evraziiskogo "stepnogo poiasa" skotovodcheskikh kul'tur: vzgliad skvoz' prizmu arkheometallurgii i radiouglerodnoi khronologii. *Arkheologiia, etnografiia i antropologiia Evrazii* **35**(3), 36-53 (2008).

Chernykh, E.N. (2017). *Nomadic Cultures in the Mega-Structure of the Eurasian World. Trans. I. Savinetskaya and P. N. Hommel.* Academic Studies Press, Boston (MA).

Chernykh (forthcoming)

Chernykh, E.N. and Lun'kov, V.Yu. (2009). Metodika rentgeno-fluorestsentnogo analiza medi i bronz v laboratorii Instituta Arkheologii. *Analiticheskiye issledovaniya laboratorii yestestvennonauchnykh metodov* **1**, 78–83.

Chu, X.L., Huo, W.G., and Zhang, X. (2002). S, C, and Pb isotopes and sources of metallogenetic elements of the Dajing Cu-polymetallic deposit in Linxi County, Inner Mongolia, China. *Acta Petrologica Sinica* **18**, 566–574.

Cleziou, S. and Berthoud. T. (1982). Early tin in the Near East: a reassessment in the light of new evidence from Afghanistan. *Expedition* **25**, 14–19.

Coghlan, H.H., Childe, V.G. and Bromehead, C.N. (1949). Ancient Mining and Metallurgy Group: Preliminary Report, Part I. *Man* **48**, 5-7.

Cooper, H.K., Duke, M.J.M., Simonetti, A. and Chen, G. (2008). Trace element and lead isotope provenance analysis of native copper in northwestern North America: results of a recent pilot study using INAA, ICP-MS, and LA-ICP-MS. *Journal of Archaeological Science* **35**, 1732-1747.

Copper Development Association (2011). *Annual data 2011. Copper: brass: bronze. Copper supply and consumption 1990–2010.* Available at: http://www.copper.org/resources/market data/pdfs/annual data.pdf (accessed 17 April 2012).

Crawford, H.E.W. (1974). The problem of tin in Mesopotamian bronzes. *World Archaeology* **6**, 242-247.

Cuénod, A. (2013). *Rethinking the Bronze-Iron Transition in Iran: Copper and Iron Metallurgy before the Achaemenid Period.* Unpublished D.Phil. thesis, University of Oxford.

Cuénod, A., Bray, P. and Pollard, A. M. (2015). The 'tin problem' in the Near East – further insights from a study of chemical datasets on copper alloys from Iran and Mesopotamia. *Iran* **LIII**, 29-48.

Cumming, G.L., Kesler, S.E. and Krstic, D. (1979). Isotopic composition of lead in Mexican mineral deposits. Economic Geology 74, 1395–407.

Damour, A. (1865). Sur la composition des haches en pierre trouvées dans les monuments celtiques et chez les tribus sauvages. *Comptes Rendues Hebdomadaires des Séances de l'Académie des Sciences* **61**, 313-321, 357-368 (1865).

Dayton, J.E. (1971). The problem of tin in the ancient world. *World Archaeology* **3**, 49-104.

Degryse, P. (ed.), 2014. *Glass Making in the Greco-Roman World.* Studies in Archaeological Sciences 4, Leuven University Press, Leuven.

de Marinis, R.C. (1997). The Eneolithic Cemetery of Remedello Sotto (BS) and the Relative and Absolute Chronology of the Copper Age in Northern Italy. *Notizie Archeologiche Bergomensi* **5**, 33–51.

Dickin, A.P. (2005). *Radiogenic isotope geology*. Cambridge, Cambridge University Press.

Dizé, M.J.J (1790), Analyse du cuivre, avec lequel les Anciens fabri-quoient leurs Médailles, les Instruments tranchans. *Observations sur la Physique, sur l'Histoire Naturelle et sur les Arts* **36**, 272-276.

Dong, L. (2012). *Xiajiadian shangcheng wenwhua kuangye yizhi de kaocha yanjiu.* Unpublished Ph.D. dissertation, University of Science and Technology Beijing.

Earl, B. and Adriaens, A. (2000). Initial experiments on arsenical bronze production. *Journal of the Minerals, Metals and Materials Society* **52**, 14–16.

Faure, G. (1986). *Principles of Isotope Geology.* John Wiley, New York (2nd edn.).

Frahm, E. and Doonan, R. (2013). The technological versus methodological revolution of portable XRF in archeology. *Journal of Archaeological Science* **40**, 1425-1434.

Gale, N. and Stos-Gale, Z., Lead and silver in the ancient Aegean. *Scientific American* **244**, 176-193 (1981).

Gale, N.H. and Stos-Gale, Z.A. (1982). Bronze Age copper sources in the Mediterranean: a new approach. *Science* **216**, 11-19.

Gilmore, G.R. and Ottaway, B.S. (1980). Micromethods for the determination of trace elements in copper-based metal artifacts. *Journal of Archaeological Science* **7**, 241-254.

Göbel, F. (1842). *Ueber den Einfluss der Chemie auf die Ermittelung der Völker der Vorzeit oder Resultate der chemischen Untersuchung metallischer Alterthümer insbesondere der in den Ostseegouvernements vorkommenden, Behuss der Ermittelung der Völker, van welchen sie abstammen.* Ferdinand Enke, Erlangen.

Godfrey, E.G. (1996), *Tin loss during bronze recycling and the archaeological evidence for the re-melting of metal in the Bronze Age Mediterranean.* Unpublished MSc Dissertation, University of Bradford.

Gosden, C. and Marshall, Y. (1999). The cultural biography of objects. *World Archaeology* **31**, 169-178.

Greenaway, F. (1962). The early development of analytical chemistry. *Endeavour* **21**, 91-97.

Hampton, D.F.G., Bennett, P., Brown, D., Lancaster, R., Rice, J.L., Sharp, A.L., Stephens, H.A. and Bidwell, H.T. (1965). Metal losses in copper-base alloys. *The British Foundryman* **58**, 225–240.

Hancock, R.G.V., Pavlish, L.A., Farquhar, R.M., Salloum, R., Fox, W.A. and Wilson, G.C. (1991). Distinguishing European trade copper and north-eastern North American native copper. *Archaeometry* **33**, 69-86.

Harris, T. (2007). Scale as Artefact: GIS, ecological fallacy, and archaeological analysis. In Lock, G. and Molyneaux, B. (eds.), *Confronting Scale in Archaeology: Issues of Theory and Practice*, pp. 39-54, Springer, New York.

Hawthorne, J.G. and Smith, C.S. (trans.) (1963). *On Divers Arts: The Treatise of Theophilus.* University of Chicago Press, Chicago.

Heginbotham, A. and Solé, V.A. (2017). Charmed PyMca, Part I: A protocol for improved inter-laboratory reproducibility in the quantitative ED-XRF analysis of copper alloys. *Archaeometry* **59**, 714–730.

Heginbotham, A., Bezur, A., Bouchard, M., Davis, J.M., Eremin, K., Frantz, J.H., Glinsman, L., Hayek, L.-A., Hook, D., Kantarelou, V., Karydas, A.G., Lee, L., Mass, J., Matsen, C., McCarthy, B., McGath, M., Shugar, A., Sirois, J., Smith, D. and Speakman, R.J. (2011). An evaluation of inter-laboratory reproducibility for quantitative XRF of historic copper alloys. In *Metal 2010: Proceedings of the Interim Meeting of the ICOM-CC Metal Working Group, October 11–15, 2010*, eds. Mardikian, P., Chemello, C., Watters, C. and Hull, P., Clemson University Press, Clemson, SC., pp. 178–188.

Heginbotham, A., Bourgarit, D., Day, J., Dorscheid, J., Godla, J., Lee, L., Pappot, A. and Robcis, D., (in press). CHARMed PyMca Part II: An Evaluation of Interlaboratory Reproducibility and Accuracy for ED-XRF analysis of Copper Alloys. *Archaeometry*.

Heginbotham, A., J. Bassett, D. Bourgarit, C. Eveleigh, L. Glinsman, D. Hook, D. Smith, R.J. Speakman, A. Shugar and R. van Langh (2015). "The Copper CHARM Set: A New Set of Certified Reference Materials for the Standardization of Quantitative X-Ray Fluorescence Analysis of Heritage Copper Alloys." *Archaeometry* **57**, 856-868.

Hodson, F.R. (1969). Searching for structure within multivariate archaeological data. *World*

Archaeology **1**, 90-105.

Hosler, D. (1995). Sound, color and meaning in the metallurgy of Ancient West Mexico. *World Archaeology* **27**, 100-115.

Howarth, P. (in prep.). *Circulation of Copper Alloys in Western Asia.* Unpublished D.Phil. thesis, University of Oxford

Hsu, Y.-K. (2016). *Dynamic Flows of Copper and Copper Alloys across the Prehistoric Eurasian Steppe from 2000 to 300 BCE.* Unpublished D.Phil. thesis, University of Oxford.

Hsu, Y.-K., Bray, P.J., Hommel, P., Pollard A.M. and Rawson, J. (2016). Tracing the flows of copper and copper alloys in the Early Iron Age societies of the eastern Eurasian steppe. *Antiquity* **90,** 357-375.

Hsu, Y.-K., Rawson, J., Pollard, A.M., Ma, Q., Luo, F., Yao P.-H. and Shen C.-C. (2018). Application of kernel density estimates to lead isotope compositions of bronzes from Ningxia, Northwest China. Archaeometry 60 128-143.

Hughes, M.J. (1979). British Middle and Late Bronze Age metalwork: some reanalyses. *Archaeometry* **21**, 195–202.

Hughes, M.J., Northover, J.P. and Staniaszek, B.E.P. (1982). Problems in the analysis of leaded bronze alloys in ancient artefacts. *Oxford Journal of Archaeology* **1**, 359-363.

Jin, Z. (1987). Sources of metals for the bronze production in the Central Plain during the late Shang period, *in Lead Isotope Archaeology in China*, ed. Z. Jin, University of Science and Technology, China, Hefei, pp. 292-302.

Jin, Z. (2008). *Zhongguo qian tongweisu kaogu.* Beijing, University of Science and Technology Press.

Jin, Z., Chase, T., Hirao, Y., Peng, S., Mabuchi, H., Miwa, K. and Zhan, K. (1994). 江西新干大洋洲商墓青铜器的铅同位素比值研究 Jiangxi Xin'gan Dayangzhou shangmu qingtongqi de qiantongweisu bizhi yanjiu, *Kaogu* 8(8), 744-747.

Jin, Z., Mabuchi, H., Chase, T., Chen, D., Miwa, K. and Hirao, Y. (1995). 广汉三星堆遗物坑青铜器的铅同位素比值研究 Guanghan Sanxingdui yiwukeng qingtongqi de qiantongweisu bizhi yanjiu. Wenwu 2(2), 80-85.

Jin, Z., Zheng, G., Hirao, Y. and Hayakawa, Y. (2001). 初期中國青銅器の鉛同位素比, in 古代東アジア青銅の流通 Kodai higashi Ajia seidō no ryūtsū, pp. 295-304, eds. Y. Hirao, J. Enomoto, and H. Takahama, 鶴山堂, Tyoko.

Jin, Z., Zhu, B., Chang, X., Xu, Z., Zhang, Q. and Tang F. (2004). 成都金沙遗址铜器研究 Chengdu Jinsha yizhi tongqi yanjiu. Wenwu (7), 76-88.

Jin, Z.Y., Liu, R., Rawson, J. and Pollard, A.M. (2017). Revisiting lead isotope data in Shang and Western Zhou bronzes. *Antiquity* **91,** 1574-1587.

Jin, Z.Y., Liu, R., Rawson, J. and Pollard, A.M. (2017). Revisiting lead isotope data in Shang and Western Zhou bronzes. Antiquity 91 1574-1587.

Junghans, S., Sangmeister, E. and Schröder, M. (1960). *Metallanalysen kupferzeitlicher und frühbronzezeitlicher Bodenfunde aus Europa.* Mann, Berlin.

Junghans, S., Sangmeister, E. and Schröder, M. (1968). *Kupfer und Bronze in der frühen Metallzeit Europas. Studien zu den Anfängen der Metallurgie.* Mann, Berlin.

Junghans, S., Sangmeister, E. and Schröder, M. (1974). *Kupfer und Bronze in der frühen Metallzeit Europas. Studien zu den Anfängen der Metallurgie.* Mann, Berlin.

Kasper, A.C., Guilherme, B.T., Berselli, B., Freitas, D., Tenório, J.A.S., Bernardes, A.M. and Veit, H.M. (2011). Printed wiring boards for mobile phones: Characterization and recycling of copper. *Waste Management* **31**, 2536-2545.

Kienlin, T.L. (2013). Copper and Bronze: Bronze Age Metalworking in Context, in H. Fokkens and A. Harding (eds.) *The Oxford Handbook of the European Bronze Age*, Oxford University Press, Oxford, pp. 414-436

Killick, D. (2001). Science, speculation and the origins of extractive metallurgy, in *Handbook of Archaeological Sciences,* eds. D.R. Brothwell and A.M. Pollard, Wiley, Chichester, pp. 483–492.

Klaproth, M.H. (1792/3). Mémoire de numismatique docimastique. *Mémoires de l'Academie Royale des Sciences et Belles-Lettres depuis l'avénement de Fréderic Guillaume II au Trône. Classe de Philosophie Expérimentale*, 97-113.

Kohl, P.L. (2007). *The Making of Bronze Age Eurasia*. Cambridge University Press, Cambridge.

Krause, R. (2003). *Studien zur kupfer- und frühbronzezeitlichen Metallurgie zwischen Karpatenbecken und Ostsee*. Marie Leidorf, Rahden/Westf.

Krause, R. and Pernicka, E. (1996). SMAP - The Stuttgart Metal Analysis Project, *Archaologisches Nachrichtenblatt* **1**, 274-291.

Krysko, W.W. (1980). Birth of Metallurgy. *Journal of Metals* **32** no 7, 43-45.

Kuijpers, M.H.G. (2013). The sound of fire, taste of copper, feel of bronze, and colours of the cast: sensory aspects of metalworking technology. In Marie Louise Stig Sørensen and Katharina Rebay-Salisbury (Eds.) *Embodied Knowledge: perspectives on belief and technology*, Oxbow Books, Oxford, pp. 137-150.

Kuijpers, M.H.G. (2018). *An archaeology of skill: metalworking skill and material specialization in early Bronze Age central Europe.* Routledge, London.

Lechtman, H. (1977). Style in technology: some early thoughts. In Lechtman, H. and Merrill, R.S. (eds.), *Material Culture: Styles, Organization, and Dynamics of the Technology*. West Publishing, St Paul, pp. 3–20.

Lee, Y.Y., Tseng, H.W., Hsiao, Y.H. and Liu, C.Y. (2009). Surface oxidation of molten Sn (Ag, Ni, In, Cu) alloys. *Journal of the Minerals, Metals and Materials Society* **61**, 52–58.

Lepsius, C.R. (1877). Les métaux dans les inscriptions Égyptiennes. *Bibliotèque de l'École des Hautes Études* **30**, 1-72.

Levey, M. (1959). *Chemistry and Chemical Technology in Ancient Mesopotamia.* Elsevier, Amsterdam.

Li, M., (1982), 殷墟金属器物成分的测定报告(一)——妇好墓铜器测定 Yinxu jinshu qiwu chengfen de ceding baogao (yi)-- Fu Hao mu tongqi ceding (Report one on the bronzes at Yinxu—the Fu Hao tomb), *Kaoguxue jikan* **2(2),** 181-193.

Li, M., Huang, S. and Ji, L. (1984). 殷墟金属器物成分的测定报告(二)--殷墟西区铜器和铅器测定 Yinxu jinshu qiwu chengfen de ceding baogao (er)-- Yinxu xiqu tongqi he qianqi ceding (Report two on the bronzes at Yinxu—the Fu Hao tomb), *Kaoguxue jikan* **4**, 328-375.

Liu Ruiliang, A.M. Pollard, Jessica Rawson and Tang Xiaojia (2017). 共性、差异与解读：运用牛津研究体系探究早商郑州与盘龙城之间的金属流通 (Revisiting the movement of metal between Zhengzhou and Panlongcheng in early Bronze Age China*). Jianghan Kaogu,* 111-121 (In Chinese).

Liu, R. (2016). *Capturing Changes: Applying the Oxford System to Further Understand the Movement of Metal in Shang China.* Unpublished D.Phil thesis, University of Oxford.

Liu, R., Pollard, A.M. and Rawson, J. (2018). Beyond ritual bronzes: multiple sources of radiogenic lead used across Chinese history. *Nature Scientific Reports* 8, Article number: 11770.

Liu, R., Bray, P., Pollard, A.M. and Hommel, P. (2015). Chemical analysis of ancient Chinese Bronzes: past, present and future. *Archaeological Research in Asia* **3**, 1–8, July 2015 (doi:10.1016/j.ara.2015.04.002).

Liu, R., Pollard, A.M., Rawson, J., Tang, X. and Zhang, C. (submitted). Panlongcheng, Zhengzhou and the Movement of Metal in Early Bronze Age China. Submitted to *Journal of World Prehistory*.

Liu, S.R., Chen, K.L., Rehren, Th., Mei, J.J., Chen, J.L., Liu, Y. and Killick, D. (2018). Lead isotope and metal source of Shang bronzes: a response to Sun *et al.*'s comments *Archaeometry* **60**, xx-yy (2018).

Liu, Y., He, Y. and Xu, G. (2007). M54及M60出土青铜器的成分分析 M54 ji M60 chutu qingtongqi de chengfen fenxi, in 安阳殷墟花园庄东地商代墓葬 *Anyang Yinxu Huayuanzhuang dongdi shangdai muzang*, ed. IA CASS, Science Press, Beijing, pp. 289-296.

Lutz, J. and Pernicka, E. (1996). Energy dispersive X-ray fluorescence analysis of ancient copper alloys: empirical values for precision and accuracy. *Archaeometry* **38**, 313–323.

McKerrell, H. and Tylecote, R.F. (1972). Working of copper-arsenic alloys in the Early Bronze Age and the effect on the determination of provenance. *Proceedings of the Prehistoric Society* **38**, 209–218.

Merkel, J.F. (1982). *Reconstruction of Bronze Age Copper Smelting, Experiments based on Archaeological Evidence from Timna, Israel.* Unpublished PhD thesis, Institute of Archaeology, University of London.

Merkl, M.B. (2011). *Bell Beaker Copper Use in Central Europe: A Distinctive Tradition?* Archaeopress, Oxford.

Mödlinger, M., Kuijpers, M.H.G., Braekmans, D. and Berger, D. (2017). Quantitative comparisons of the color of CuAs, CuSn, CuNi, and CuSb alloys. *Journal of Archaeological Science* **88**, 14-23.

Muhly, J.D. (1985). Sources of tin and the beginnings of bronze metallurgy. *American Journal of Archaeology* **89**, 275-291.

Muhly, J.D. (1993). Ancient metallurgy in the USSR: The early metal age. E. N. Chernykh, 1992, *Geoarchaeology* **8**, 534–553.

Needham, S.P. (1996). Chronology and periodisation in the British Bronze Age. In *Absolute Chronology: Archaeological Europe 2500-500 B.C.,* ed. K. Randsborg, Munksgaard, København, pp. 121-140.

Needham, S.P. (1998). Modelling the flow of metal in the Bronze Age. In *L'Atelier du Bronzier en Europe du XX au VIII Siecle Avant Notre Era*, eds. Mordant, C., Pernot, M. and Rychner, V., Actes du colloque international Bronze '96, CTHS, Paris, pp. 285-307.

Neugebauer, J.-W. (1991). *Die Nekropole F von Gememeinlebarn, Niederösterreich: Untersuchungen zu den Bestattungssitten und zum Grabraub in der ausgehenden Frühbronzezeit in Niederösterreich südlich der Donau zwischen Enns und Wienerwald.* Unpublished PhD thesis, Universität Wien.

Nezafati, N., Pernicka, E. and Momenzadeh, M. (2009). Introduction of the Deh Hosein ancient tin-copper mine, western Iran: Evidence from geology, archaeology, geochemistry and lead isotope data. *Tüba-Ar* **12**, 223–235.

Northover, J.P. (1980). Appendix. The analyses of Welsh Bronze Age metalwork. In *Guide Catalogue of the Bronze Age Collections,* Savory, H.N., National Museum of Wales, Cardiff, pp. 229-243.

Northover, J.P. and Rychner, V. (1998). Bronze analysis, experience of a comparative programme. In *L'Atelier du Bronzier en Europe du XX au VIII Siecle Avant Notre Era*, eds. Mordant, C., Pernot, M. and Rychner, V., Actes du colloque international Bronze '96, CTHS, Paris, pp. 19-40.

Northover, P. (1999). The earliest metalworking in southern Britain. In *The Beginnings of Metallurgy,* eds. Hauptman, A., Pernicka, E., Rehren, T. and Yalcin, U., Deutches Bergbau-

Museum, Bochum, pp. 211-226.

O'Brien, W. (2015). *Prehistoric Copper Mining in Europe*. Oxford University Press, Oxford.

Oddy, W.A. (1986). The touchstone: the oldest colorimetric method of analysis. *Endeavour* **10**, 164–166.

Okunev, A.I. (1960). *Povedenie nekotorikh redkikh i rasseyannykh elementov v protsesse metallurgicheskoy pererabotki mednykh rud I kontsentratov Okunev (*Behavior of some rare and dispersed elements in the process of metallurgical processing of copper ores and concentrates*)*, Nauka, Moscow.

Oldroyd, D.R. (1973). Some eighteenth century methods for the chemical analysis of minerals. *Journal of Chemical Education* **50**, 337-340.

Olin J.S. and Wertime, T.A. (eds.). *The search for Ancient Tin*. Smithsonian Institution Press, Washington, D.C. (1978).

Ottaway, B.S. (1978). *Aspects of the Earliest Copper Metallurgy in the Northern Sub-Alpine Area and its Cultural setting*. PhD Dissertation, University of Edinburgh.

Ottaway, B.S. (1982). *Earliest Copper Artefacts of the North Alpine Region: Their analysis and evaluation*. Universität Bern, Bern.

Ottaway, B.S. (2001). Innovation, production and specialization in early prehistoric copper metallurgy. *European Journal of Archaeology* **4**, 87-112.

Otto, H. and Witter, W. (1952). *Handbuch der ältesten vorgeschichtlichen Metallurgie in Mitteleuropa*. Barth, Leipzig.

Partington J.R. (1935). *Origins and Development of Applied Chemistry*. Longmans, Green and Co., London (reprinted 1975).

Partington, J.R. (1936). The discovery of bronze. *Scientia, rivista di scienza* **60,** 197-204.

Peng, Z., Liu, Y., Liu, S. and Hua, J. (1999). 赣鄂豫地区商代青铜器和部分铜铅矿料来源的初探 Gan'eyu diqu shangdai qingtongqi he bufen tongqian kuangliao laiyuan de chutan (The preliminary exploration of the source of bronzes and some lead material in Jiangxi, Hubei and Anhui), *Ziran kexueshi yanjiu* **18**(3), 241-249.

Peng, Z., Wang, Z., Sun, W., Liu, S. and Chen, X. (2001). 盘龙城商代青铜器铅同位素示踪研究 Panlongcheng Shangdai qingtongqi qiantongweisu shizong yanjiu (Using lead isotopes to source Panlongcheng bronzes), in 盘龙城1963-1994年考古发掘报告 Panlongcheng 1963-1994nian kaogu fajue baogao (The excavation report of Panlongcheng in the year 1963-1994), pp. 552-558, ed. 湖北省文物考古研究所 Hubei Institute of Archaeology, Wenwu press, Beijing.

Pernicka, E. (1995). Gewinnung und Verbeitung der Metalle in prähistorischer Zeit. *Jahrbuch des Römische-Germanischen Zentralmuseum* **37**, 21-129.

Pernicka, E. (1999). Trace element fingerprinting of ancient copper: a guide to technology or provenance? In *Metals in Antiquity,* eds. Young, S.M.M., Pollard, A.M., Budd, P. and Ixer, R.A., BAR International Series 792, Archaeopress, Oxford, pp. 163-171.

Pernicka, E. (2011). A short history of provenance analysis of archaeological metal objects. In *I Bronze del Garda*, ed. Aspes A., Museo Civica di Storia Naturale, Verona, pp. 27-37.

Pernicka, E., (1986). Provenance determination of metal artifacts: methodological considerations. *Nuclear Instruments and Methods in Physics Research Section B: Beam Interactions with Materials and Atoms* **14**, 24–29.

Pernicka, E., (1990). Entstehung und Ausbreitung der Metallurgie in prähistorischer Zeit, *Jahrbuch der Römanisch-Germanische Zentralmuseum* **37**, 21-129.

Pernicka, E., Begemann, F.S., Schmitt-Strecker, S., Todorova, H. and Kuleff, I. (1997). Prehistoric copper in Bulgaria. *Eurasia Antiqua* **3**, 41-180.

Perucchetti, L. (2017). *Physical barriers, cultural connections: A reconsideration of the metal flow at the beginning of the Metal Age in the Alps.* Archaeopress, Oxford.

Perucchetti, L., Bray, P., Dolfini, A. and Pollard, A.M. (2015). Physical barriers, cultural connections: prehistoric metallurgy in the Alpine region. *European Journal of Archaeology* **18**, 599–632.

Pickles, C.A. (1998). Selective oxidation of copper from liquid copper-silver alloys. *Metallurgical and Materials Transactions B* **29**, 39–51.

Pittioni, R. (1957). *Urzeitlicher Bergbau auf Kupfererz und Spurenanalyse: Beiträge zum Problem der Relation Lagerstätte-Fertigobjekt.* Dueticke, Wien.

Pittioni, R. (1960). Metallurgical analysis of archaeological materials. II. In *The Application of Quantitative Methods in Archaeology,* eds. Heizer, R.F. and Cook, S.F. (Viking Fund publications in anthropology, No. 28), Quadrangle Books, Chicago, pp. 21-24 (Discussion, pp. 25-33).

Pollard, A.M. (1983). A critical study of multivariate methods as applied to provenance data. In *Proceedings of the 22nd Symposium on Archaeometry, University of Bradford, March 30th - April 3rd, 1982,* eds. Aspinall, A. and Warren, S.E., University of Bradford, pp. 56-66.

Pollard, A.M. (2009). 'What a long strange trip it's been': lead isotopes in archaeology. In *From Mine to Microscope,* eds. Shortland A.J., Freestone, I.C. and Rehren, T., Oxbow Books, Oxford, pp. 181-189.

Pollard, A.M. (2013). From bells to cannon - the beginnings of archaeological chemistry in the Eighteenth Century. *Oxford Journal of Archaeology* **32,** 333-339.

Pollard, A.M. and Bray, P.J. (2015). A new method for combining lead isotope and lead abundance data to characterise archaeological copper alloys. *Archaeometry* **57** 996–1008.

Pollard, A.M., Batt, C., Stern, B. and Young, S.M.M. (2007). *Analytical Chemistry in Archaeology.* Cambridge, Cambridge University Press.

Pollard, A.M., Bray, P., Gosden, C., Wilson, A. and Hamerow, H. (2015). Characterising copper-based metals in Britain in the first millennium AD: A preliminary quantification of metal flow and recycling. *Antiquity* **89,** 697–713.

Pollard, A.M., Bray, P., Hommel, P., Hsu, Y.-K., Liu, R. and Rawson, J. (2017a). Bronze Age metal circulation in China. *Antiquity* **91**, 674-687.

Pollard, A.M., Bray, P.J. and Gosden, C. (2014). Is there something missing in scientific provenance studies of prehistoric artefacts? *Antiquity* **88,** 625-631.

Pollard, A.M., Heron, C. and Armitage, R.A. (2017b). *Archaeological Chemistry.* Royal Society of Chemistry, Cambridge (3rd revised edition).

Pollard, A.M., Liu, R. and Rawson, J. (in prep.), Every cloud has a silver lining: using silver concentration to identify the number of sources of lead used in Shang Dynasty Bronzes. Submitted to *Antiquity.*

Pollard, A.M., Liu, R., Rawson, J. and Tang, X. (2018b). From alloy composition to alloying practice: Chinese bronzes. *Archaeometry* **60**, xx-yy.

Pollard, A.M., Rawson, J. and Liu, R. (2018). Some recently rediscovered analyses of Chinese bronzes from Oxford. *Archaeometry* **60**, 118-127.

Preuschen, E. and Pittioni, R. (1937). Untersuchungen im Bergbaugebiete Kelchalpe bei Kitzbühel, Tirol: 1. Bericht über die Arbeiten 1931-1936 zur Uregeschichte des Kupferbergwesens in Tirol. *Mitteilungen der Prähistorischen Kommission* **3**, 1-159.

Radivojević, M., Pendić, J., Srejić, A., Korać, M., Davey, C., Benzonelli, A., Martinon-Torres, M., Jovanović, N. and Kemerović, Z. (2018). Experimental design of the Cu-As-Sn ternary colour diagram. *Journal of Archaeological Science* **90**, 106-119.

Radivojević, M., Rehren, Th., Kuzmanović-Cvetković, T.J., Jovanović, M. and Northover, J.P. (2013). Tainted ores and the rise of tin bronze metallurgy, c. 6500 years ago. *Antiquity* **87,** 1030-1045.

Radivojević, M., Roberts, B.W., Pernicka, E., Stos-Gale, Z., Martinón-Torres, M., Rehren, Th., Bray, P., Brandheim, D., Ling, J., Mei, J., Vandkilde, H., Kristiansen, K., Shennan, S.J. and Broodbank, C. (2018). The provenance, use, and circulation of metals in the European Bronze Age: The state of debate. *Journal of Archaeological Research*, 1–55.

Ravasi, T. and Barbaglio, F. (2008). Merci e Persone sui Fiumi. Le imbarcazioni monossili conservate presso in Museo Civico di Crema . In M. Baioni and C. Fredella (eds.) *Archaeotrade. Antichi Commerci in Lombardia Orientale*, pp. 37-61, ET. Milano.

Rawson, J. (1990). *Western Zhou ritual bronzes from the Arthur M. Sackler Collections*. Arthur M. Sackler Foundation, Washington, D.C.

Rawson, J. (2015). Steppe weapons in Ancient China: China and the role of hand-to-hand combat. *National Palace Museum Quarterly* **33**, 37-96.

Rehren, T., Boscher, L. and Pernicka, E. (2012). Large scale smelting of speiss and arsenical copper at Early Bronze Age Arisman, Iran. *Journal of Archaeological Science* **39**, 1717-1727.

Roberts, B.W., Thornton, C.P. and Pigott, V.C. (2009). Development of metallurgy in Eurasia. *Antiquity* **83,** 1012-1022.

Russell, R.D. and Farquhar, R.M. (1960). *Lead Isotopes in Geology*. Interscience Publishers, New York.

Sabatini, B. (2017). *Chemical Composition, Thermodynamics, and Recycling: the Beginnings of Predictive Behavioral Modeling for Ancient Copper-based Systems*. Unpublished DPhil thesis, University of Oxford.

Scaife, B., Budd, P., McDonnell, J.G. and Pollard, A.M. (1999). Lead isotope analysis, oxhide ingots and the presentation of scientific data in archaeology. In *Metals in Antiquity*, eds. Young, S.M.M., Pollard, A.M., Budd, P. and Ixer, R.A., BAR International Series 792, Archaeopress, Oxford, pp. 122-133.

Scaife, B., Budd, P., McDonnell, J.G., Pollard, A.M. and Thomas, R.G. (1996). A new statistical technique for interpreting lead isotope analysis data. In Demirci, Ş., Özer, A.M. and Summers, G.D. (eds.) *Archaeometry 94. Proceedings of the 29th International Symposium on Archaeometry*. Tübitak: Ankara, Turkey, pp. 301-307.

Schubert, F. and Schubert, E. (1967). Spektralanalytische Untersuchungenvon Hort- und Einzelfunden der Periode BIII. In *Bronzefunde des Karpatenbeckens: Depotfundhorizonte von Hajdúsámson und Kosziderpadlás,* Mozsolics, A., Akadémiai Kiadó, Budapest, pp. 185-223.

Shackley. M.S. (ed.) (2011). *X-ray Fluorescence Spectrometry (XRF) in Geoarchaeology*. Springer, New York.

Shugar, A.N. and Mass, J. (eds.) (2013). *Handheld XRF for Art and Archaeology*. Studies in Archaeological Sciences 3. Leuven University Press, Leuven.

Sisco, A.G. and Smith, C.S. (trans.) (1949). *Bergwerk- und Probierbüchlein*. American Institute of Mining and Metallurgical Engineers, New York.

Sisco, A.G. and Smith, C.S. (trans.) (1951). *Lazarus Ercker's Treatise on Ores and Assaying*. University of Chicago Press, Chicago.

Slater, E.A. and Charles, J.A. (1970). Archaeological classification by metal analysis. *Antiquity* **XLIV**, 207-213.

Smith, D.M. (1933). *Metallurgical Analysis by the Spectrograph*. Research Monograph No. 2, British Non-Ferrous Metals Association, London.

Smith, E.A. (1935). The origin of metal alloying. *Metal Industry (London)* **46**, 375-377.

Sneath, P.H.A. and Sokal, R.R. (1973). *Numerical Taxonomy: the Principles and Practice of Numerical Classification*. Freeman, San Francisco.

Solecki, R. (1969). A copper mineral pendant from northern Iraq. *Antiquity* **43**, 311-314.

Speakman, R.J. and Shackley, S. (2013). Silo science and portable XRF in archaeology: a response to Frahm. *Journal of Archaeological Science* **40**, 1435–1443.

Spriggs, M. (1989). The dating of the Island Southeast Asian Neolithic: an attempt at chronometric hygiene and linguistic correlation. *Antiquity* **63**, 587-613.

Staniaszek, B.E.P. and Northover, J.P. (1983). The properties of leaded bronze alloys. In *Proceedings of the 22nd International Symposium on Archaeometry (Bradford),* eds. Aspinall, A. and Warren, S.E., School of Physics and Archaeological Sciences, University of Bradford, pp. 262-272.

Stöllner, T. and Samašev, Z. (2013). *Unbekanntes Kasachstan. Archäologie im Herzen Asiens*. Deutschen Bergbau-Musem, Bochum (2 vols.).

Struve, H. (1866). Analyse verschneidener antiker Bronzen und Eisen aus der Abakan- und Jenissei-Steppe in Siberien. *Bulletin de l'Académie Impériale des Sciences de St.-Petersbourg* **9**, 282-290.

Sullivan, W.K. (1873). The sources and composition of the ancient bronzes of Europe. In O'Curry, E., *On the Manners and Customs of the Ancient Irish. Vol. I. Introduction*, pp. ccccvii – ccccxxx, Williams and Norgate, London.

Sun, S., Han, R., Chen, Zhai, T., Ban, B. and Tian, K. (2001). 盘龙城出土青铜器的铅同位素比测定报告 Panlongcheng chutu qingtongqi de qiantongweisubi ceding baogao (Report on the lead isotopic ratios of bronzes at Panlongcheng), in 盘龙城: 1963-1994年考古发掘报告 Panlongcheng: 1963-1994 nian kaogu fajue baogao (The excavation report of Panlongcheng excavation in the year 1963-1994), pp. 545-551, ed. 湖北省文物考古研究所 Hubei Institute of Archaeology, Wenwu press, Beijing.

Sun, W., Zhang, L., Guo, J., Li, C., Jiang, Y., Zartman, R.E. and Zhang, Z. (2016). Origin of the mysterious Yin-Shang bronzes in China indicated by lead isotopes. Nature Scientific Reports 6, 1-9.

Tanahashi, M., Fujinaga, T., Su, Z.J., Takeda, K., Sohn, H.Y. and Yamauchi, C. (2005). Effects of coexisting oxygen and antimony in molten copper on rate of arsenic elimination from the copper phase by the use of Na_2CO_3 slag. *Materials Transactions* **46**, 2180–2189.

Teppo, O., Niemela, J. and Taskinen, P. (1991). The copper-lead phase diagram. *Thermochimica Acta* **185**, 155-169.

Thompson, F.C. (1958). The early metallurgy of copper and bronze. *Man* **58**, 1-7.

Thornton, C.P. and Roberts, B.W. (eds.) (2014). *Archaeometallurgy in Global Perspective: Methods and Syntheses*. Springer, New York.

Tian, J.,郑州地区出土二里岗期铜器研究 Zhengzhou diqu chutu Erligangqi tongqi yanjiu (The study of the Erligang bronzes at Zhengzhou), PhD thesis, University of Science and Technology China, Hefei (2013).

Tong, E. (1987). 试论我国从东北至西南的边地半月形文化传播带 Shilun woguo cong dongbei zhi xinan de biandi banyuexing wenhua chuanbodai (The crescent of China from southwest to northeast borderlands). In 中国西南民族考古论文集 *Zhongguo xinan minzu kaogu lunwenji*, ed. Tong, E., Wenwu press, Beijing, pp. 252-278.

Tylecote, R.F. (1970). The composition of metal artifacts: a guide to provenance? *Antiquity* **XLIV**, 19-25.

van der Stok-Nienhuis, J. (2017). *Artefact biography 2.0: the information value of corroded archaeological bronzes*. Technische Universiteit Delft, Netherlands.

Vauquelin, L.N. (1799). Réflections sur l'analyse des pierres en general, et résultats de plusiers de ces analyses faites au laboratoire de l'école des mines depuis quelques mois. *Annals de Chimie* **30**, 66-106.

von Bibra, E.F. (1869). *Die Bronzen und Kupferlegirung der alten und ältesten Völker, mit Rücksichtnahme aur jene der Neuzeit*. Ferdinand Enke, Erlangen.

Wang, H., Cowell, M., Cribb, J. and Bowman, S. (eds.) (2005). *Metallurgical Analysis of Chinese Coins at the British Museum*. British Museum Research Publication Number 152, British Museum, London.

Wang, H., Jin, Z., Li, G., Tian, J., Li, R. and Zhao, C. 2008. 城洋地区出土部分青铜器的科学分析 Chengyang diqu chutu bufen qingtongqi de kexue fenxi. Xibu kagu (3), 319-321.

Waterbolk, H.T. and Butler, J.J. (1965). Comments on the use of metallurgical analysis in prehistoric studies. *Helinium* **V**, 227-251.

Willett, F. and Sayre, E.V. (2000). The elemental composition of Benin memorial heads. *Archaeometry* **42**, 159-188.

Wilson, L. and Pollard, A.M. (2001). The provenance hypothesis. In *Handbook of Archaeological Sciences*, eds. Brothwell, D.R. and Pollard, A.M., John Wiley and Sons, Chichester, pp. 507-517.

Xu, W., Wang, L., Li, H. and Guo, X. (2005) 中条山铜矿床同位素地球化学研究 Zhongtiaoshan tongkuangchuang tongweisu diqiu huaxue yanjiu. Acta Geoscientica Sinica 26(增刊), 130-133.

Yang, J. (2015). *Chifeng diqu qingtong shidai wanqi tongqi de kexue fenxi yanjiu*, Unpublished Ph.D. thesis, University of Science and Technology Beijing.

Young, M.L., Casadio, F., Marvin, J., Chase, W.T. and Dunand, D.C. (2010). An ancient Chinese bronze fragment re-examined after 50 years: contributions from modern and traditional techniques. *Archaeometry* **52**, 1015-1043.

Zhao, C. (2004). 安阳殷墟出土青铜器的化学成分分析与研究 Anyang Yinxu chutu qingtongqi de huaxue chengfen fenxi yu yanjiu, 考古学集刊 *Kaoguxue jikan* **15**(15), 243-268.

Zhu, H., Gao, J. and Zhang, D. (2006). 秦岭地区首次发现含放射性成因异常铅的铜矿床 Qinling diqu shouci faxian han fangshexing chengyin yichangqian de tongkuangchuang. Mineral Resources and Geology 20(4-5), 461-464.

Bibliography of Sources of Chemical and Isotopic Data Used in the FLAME Database

Abesadze, C., 1969. *Proizvodstovo Metalla v Zakavkaze v III tysjacheletii do. n.e (Kuro-Araksskaja kult'tura).* Izdatel'tvo akademija nauk Gruzinskoj, Tbilissi.

Abesadze, T.N. and Bakhtadze, R.A., 1961. Klad Pozdnebronzovoy Epokhi iz c. Ude (Yuzhnaya Gruziya). *Sovetskaia arkheologiia* **3**, 166–177.

Adams, F., 2003, Microanalytical metal technology study of ancient near eastern bronzes from Tell Beydar. *Archaeometry* **45**, 579-590.

Adams, R.B., 2002, Early Bronze Age Metallurgy: a newly discovered copper manufactory in southern Jordan. *Antiquity* **76**, 425-437.

Agapov, S.A. (ed), 2010. *Khvalynskie Eneoliticheskie Mogil'niki I Khvalynskaya Eneolitichaekaya Kul'tura*. SROO IEKA, 'Povolzh'e' Samara.

Agapov, S.A., 1990. *Metall stepnoy zony Evrazii v kontse Bronzogo Veka.* Unpublished Dissertation (Kandidat Istoricheskiy Nauk), Moscow.

Akhundov, T.I., 2014. At the Beginning Caucasian Metallurgy. In Narimanishvili G.K., Kvačaże, M., Puturidze, M. and Šanšašvili, N (eds.), *Problems of Early Metal Age Archaeology of Caucasus and Anatolia.* GNM Press, Tbilisi, pp. 11–16.

Alekseyev, V.A. and Kuznetsova, E.F., 1983. Kenkazgan: drevniy mednyy rudnik v Tsentral'nom Kazakhstane. *Rossiyskaya Arkheologiya* **2**, 203–212.

Angelini, I. and Artioli, G., 2006. Studio archeometallurgico di noduli metallici da Santa Caterina – Tredossi (Castelverde, CR), in *La Terramara Di Santa Caterina Tredossi (Cremona)*, pp. 135–140.

Angelini, I. and Artioli, G., 2007. Le armi eneolitiche di Spessa (Cologna Veneta) e di Gambarella (Povegliano): indagini chimiche e tessiturali. *Bollettino Del Museo Civico di Storia Naturale di Verona* **31**, 51–61.

Angelini, I. and Salzani, L., 2005. Le armi della necropoli di Olmo di Nogara: analisi chimiche, metallografiche e microstrutturali, in *La Necropoli dell'Età Del Bronzo all'Olmo Di Nogara (Verona)*, Verona, pp. 515–527.

Angelini, I., 2004. Studio archeometallurgico di un ornamento a spirale dal Riparo di Peri (Verona). *Padusa* **XL**, 39–50.

Angelini, I., 2007. Indagini archeometriche dell'ascia di Caprino Veronese. *Quaderni Culturali Caprinesi* **2**, 34–41.

Angelini, I., Polla, A., Perucchetti, L., de Marinis, R.C. and Artioli, G., 2012. Analisi archeometallurgiche di asce e pugbnali del bronzo antico-bronzo medio provenienti dall'Italia settentrionale, in Riccardi, M.P. and Basso, E (eds.), *Atti del Vi Congresso Nazionale di Archeometria Scienza e Beni Culturali. Pavia.* (CD-ROM).

Armbruster, B., 2006. Steingerate des bronzezeitlichen Metallhandwerks. *Ethnographisch-archaolische Zeitschrift* **47**, 163-191.

Askarov, A.A. and Ruzanov, V.D., 1992. Rezultaty Spektral'nykh Issledovaniya Metalla iz Poseleniya Kuchuktepa. *Istoriya Materialnoy Kul'tury Uzbekistana* **26**, 211–25.

Askarov, A.A., Abdurazakov, A.A., Bogdanova-Berezovskaya, N. V., and Ruzanov, V.D., 1975. Khimicheskiy Sostav Metallicheskikh Predmetov iz Poseleniya Sapallitepa. *Istoriya Materialnoy Kul'tury Uzbekistana* **12**, 72–84.

Bader, O.N., 1964. *Drevneyshie Metallurgiya Priural'ya*, Nauka, Moscow.

Bagley, R.W., 1987. *Shang Ritual Bronzes in the Arthur M. Sackler Collections*, Arthur M. Sackler Foundation, Washington D.C.

Bai, C.B., Qi, Y. and Yang, Q.Y. 2006. Baoji bufen xizhou qingtongqi de nengpu fenxi yanjiu. In School of Cultural Heritage of Northwest University, X'ian (eds.), *Wenwu baohu yu keji kaogu*, pp. 21-25, Sanqin, Xi'an.

Balthazar, J.W., 1990. *Copper and Bronze Working in Early Through Middle Bronze Age Cyprus*. Paul Åströms Förlag, Sweden.

Barker, G., 1971. Analyses of samples from the Pigorini Museum. *Bullettino di Palentologia Italiana* **80**, 183–212.

Bartseva, T., 1984. Khimicheskiy sostav nakonechnikov kopiy severnogo Kavkaza VIII–VII do NE. *Kratkie soobshchenii͡a* **184**, 42–49.

Bartseva, T.B., 1978. O khimicheskom sostave bronzogog kotla iz s. Gaydary. *Sovetskai͡a arkheologii͡a* **1978.2**, 256–257.

Bartseva, T.B., 1980. O Khimicheskim Sostave Metalla Navershiy Skifskogo Vremeni. *Sovetskai͡a arkheologii͡a* **1980.3**, 77–91.

Bartseva, T.B., 1981. *Tsvetnaya Metalloobrabotka Skifskogo Vremeni: Lesostepnoe Dneprovskoe Levoberezh'e*. Nauka, Moscow.

Bartseva, T.B., 1983. Khimichskiy Sostav Izdeliy Antichnogo Importa Naydennykh v Srednem Podneprov'e. *Sovetskai͡a arkheologii͡a 1983.4*, 70–82.

Bartseva, T.B., 1985. Khimicheskiy sostav nakonechnikov kopiy severnogo Kavkaza VIII–VII do N.E. *Kratkie Soobshchenii͡a Instituta Arkheologii* **184**, 42–49,

Bartseva, T.B., 1988. Bronzovye kinzhaly Serzhen'-Yurta. *Kratkie soobshchenii͡a* **194**, 23–29.

Begemann, F. and Schmitt-Strecker, S., 2009. Uber das Fruhe Kupfer Mesopotamiens. *Iranica Antiqua* **44**, 1-45 (2009).

Begemann, F., Hauptmann, A., Palmierei, A., and Schmitt-Strecker, S., 2002. Chemical composition and lead isotopy of metal objects from the "Royal" tomb and other related finds at Arslantepe, Eastern Anatolia. *Palaeorient* **28**, 43–69.

Begemann, F., Hauptmann, A., Schmitt-Strecker, S. and Weisgerber, G., 2010. Lead isotope and chemical signature of copper from Oman and its occurrence in Mesopotamia and sites on the Arabian Gulf coast. *Arabian Archaeology and Epigraphy* **21**, 135-169 (2010).

Begemann, F., Heirink, E., Overlaet, B., Schmitt-strecker, S. and Tallon, F., 2008. An archaeo-metallurgical study of the Early and Middle Bronze Age in Luristan, Iran. *Iranica Antiqua* **43**, 1-66 (2008).

Begemann, F., Pernicka, E. and Schmitt-Strecker, S., 1994. Metal Finds from Ilipinar and the advent of arsenical copper. *Anatolica* **XX**, 203–212.

Begemann, F., Schmitt-Strecker, S. and Pernicka, E., 2003. On the comparison and provenance of metal finds from Besiktepe, in Wagner, G.A., Pernicka, E., Uerpmann, H.-P. (eds.), *Troia and the Troad: Scientific Approaches*. Springer, Berlin, pp. 173–202.

Bennett, A., 1988. The contribution of metallurgical studies to South East Asian archaeology. *World Archaeology* **20**, 329-351.

Berthoud, T., 1980. *Contribution à l'étude de la métallurgie de Suse aux IVème et IIIème millénaires: analyse des éléments-traces par spectrométrie d'émission dans l'ultra-violet et spectrométrie de masse à étincelles.* CEN, Paris.

Berzero, A., Caramella Crespi, V., Genova, N., Meloni, S., Oddone, M. and Pearce, M., 1991. Indagine chimica sul ripostiglio di Pieve Albignola (Pavia), in *Cataloghi dei Civici musei di Pavia. Assessorato alla cultura, Civici musei di Pavia, Pavia*, pp. 163–170.

Besse, M., 2011. Provenance of Early Bronze Age metal artefacts in Western Switzerland using elemental and lead isotopic compositions and their possible relation with copper minerals of the nearby Valais. *Journal of Archaeological Science* **38**, 1221–1233.

Birmingham, J., 1963. Iranian bronzes in the Nicholson Museum, University of Sydney. *Iran* **1,** 71–82.

Blin-Stoyle, A.E., 1959. Chemical composition of the bronzes. *Archaeometry* **2**, 1–18.

Boardman, J., 1961. *The Cretan Collection in Oxford*. Oxford, Clarendon Press.

Bobrov, V.V., 1976. The manufacturing technology of bronze figures of deer. *Proceedings of the archaeological research laboratory* **7**, 81-86.

Bobrov, V.V., S.V. Kuz'minykh, and Teneyshvili, T.O., 1997. *Drevnyaya Metallurgiya Srednego Yeniseya (Lugavskaya Kul'tura).* Kuzbassvuzizdat, Kemerovo.

Bogdanova-Berezovskaya, V.I., 1962. Khimicheskiy sostav metallicheskikh izdeliy Fergany Epokhi Bronzy i Zheleza. *Materialy i issledovaniîâ po arkheologii SSSR* **118**, 219–230.

Bogdanova-Berezovskaya, V.I., 1963. Khimicheskiy sostav metallicheskikh predmetov iz Minusinskoy kotloviny, [Chemical composition of metal objects in the Minusinsk basin], in S.I. Rudenko (ed.), *New methods in archaeological research,* Akademii nauk SSSR, Leningrad, pp. 115-158.

Borodovskiy, A.P., 2012. Predvaritel'ie Rezultaty Issledovanii Sostav Metalla Izdelii Iyusskogo Klada (Respublika Khakassiya). In Borodovskiy, A.P., Tishkin, A.A. and Khavrin, S.V. (eds), *Problemy Arkheologii, Etnografii, Antropologii Sibir I Sopredel'nykh Territoriy,* pp. 175–179, Institut Arkheologii i Etnographii SORAN, Novosibirsk.

Briard, J. and Maréchal, J.R., 1958. Etude technique d'objets métalliques du Chalcolithique et de l'Age du Bronze de Bretagne. *Bulletin de la Société préhistorique française* **55**, 422-430.

Brileva, O.A., 2009. Noveyshie Issledovaniya Pamyatnikov Epokhi Bronzy Kavkazkoy Arkheologicheskoy Ekspeditsiya Gosudarstvennogo Muzeya Narodov Vostoka. In *Pyataya Kubanskaya Arkheologixheskaya Konferentsiya*. Kuban State University Press, Krasnodar, pp.18–20.

Britton, D., 1961. A study of the composition of Wessex Culture bronzes. *Archaeometry* **4**, 39–52.

Britton, D., 1963. Traditions of metal-working in the later Neolithic and Early Bronze Age of Britain: Part I. *Proceedings of the Prehistoric Society (New Series)* **29,** 258–325.

Brovender, Yu.M., 2009. Metall Poseleniya Srubnoy Obshchnosti Chervone Ozero-3 v Tsentral'nom Donbasse. In Yu.M. Brovender and O.A. Kovalenko (eds) *Problemy Arkheologii Podniprov'ya*, pp. 75–79, Vidavnitstvo DNU, Dniepropetrovsk.

Budd, P., 1991. A metallographic investigation of Eneolithic arsenical copper artefacts from Mondsee, Austria. *Historical Metallurgy* **25**, 99–108.

Bunimovitz, S., 2000. Metal artefacts, in Kochavi, M. (ed.), *Aphek-Antipatris I: Excavations of Areas A and B Theb 1972 - 1976 Seasons.* Emery and Claire Yass Publications in Archaeology, Tel Aviv.

Bunker, E.C., Kawami, T.S., Linduff, K.M. and Wu, E., 1997. *Ancient Bronzes of the Eastern Eurasian Steppes from the Arthur M. Sackler collections*. Arthur M. Sackler Foundation, New York.

Buresta, E, Gaggi, C., Giardino, C., Moroni Lanfredini A., Nicolardi, V. and G. Protano, 2006. Indagini archeometallurgiche su reperti preistorici dalla Val di Chiana: lo sfruttamento dei giacimenti toscani nelle prime fasi delle età dei metalli. *Rivista di scienze preistoriche*: **LVI**, 273-292.

Burov, V.M., 1983. O Nizhem Khronologicheskom Predele Lebyazhskoy Kul'tury. *Sovetskai͡a arkheologii͡a* **1983.2**, 34–50.

Burton Brown, T., 1951. *Excavations in Azarbaijan, 1948*. John Murray, London.

Bushmakin, A.F., 2002. Metallicheskiye predmety iz kurgana 25 Bol'shekaraganskogo mogil'nika, in Zdanovich, D.G. (ed) *Arkaim: Necropolis*, Kamennyy Poyas, Chelyabinsk, pp.132-143.

Caley, E.R., 1972. Chemical examination of metal artifacts from Afghanistan, in Dupree, L. (ed.), *Prehistoric Research in Afghanistan (1959-1966)*. The American Philosophical Society, Philadelphia, pp. 44–50.

Cambi, L. and Elli, M., 1960. Relitto di fonderia terramaricola. *Studi Etruschi* **XXVIII**, 421–435.

Cambi, L., 1959. Ricerche chimico metallurgiche su leghe cuprifere di oggetti ornamentali preistorici dell'Italia Centrale e Settentrionale. *Studi Etruschi* **XXVII**, 191–198.

Cambi, L., 1960. Problemi della Metallurgia Etrusca, in *Atti Del III Convegno Di Studi Etruschi*. Presented at the Convegno di Studi Etruschi, Chiusi, pp. 64–80.

Caneva, C. and Palmieri, A., 1983. Metalwork at Arslantepe in Late Chalcolithic and Early Bronze I: The evidence from metal analysis. *Origini Rivista di Prehistoria e Protostoria delle Civiltà Antiche* **12**, 637–654.

Cao, D., 2014. *The Loess Highland in a Trading Network (1300-1050BC)*. PhD thesis, Princeton University.

Cattin, F., Guénette-Beck, B., Curdy, P., Meisser, N., Ansermet, S., Hofmann, B., Kündig, R. and Coghlan, H.H., 1968. A metallurgical examination of some prehistoric flat and flanged axes, *Archeologia Austriaca* **44,** 61-82.

Cattin, F., Villa, I.M. and Besse, M., 2009. Copper supply during the Final Neolithic at the Saint-Blaise/Bains des Dames site (Neuchâtel, Switzerland). *Archaeological and Anthropological Science* **1**, 161–176.

Ceng, L., Xia, F., Xiao, M.L. and Shang, Z. 1990. Sunan diqu gudai qingtongqi hejin chengfen de ceding. *Wenwu* **9**, 37-47.

Chase, T.C., 1996-97. Lead isotope ratio analysis of Chinese bronzes - examples from the Freer Gallery of Art and Arthur M. Sackler collections, in Bulbeck, F.D. and Barnard, N. (eds.) *Ancient Chinese and Southeast Asian Bronze Age Cultures,* SMC Publishing Inc., Taipei, pp. 499-534.

Chase, T.C., and Douglas, J.G., 1997. Technical studies and metal compositional analysis of bronzes from the Eastern Eurasian Steppes, in Bunker, E.C. (ed.) *Ancient Bronzes of the Eastern Eurasian Steppes from the Arthur M. Sackler Collections*, The Arthur M. Sackler Foundation, New York, pp. 306-418.

Chen, J., Sun, S., Han, R., Chen, T., Zhai, T. Ban, B. and Tian, K., 2001, 盘龙城遗址出土铜器的微量元素分析报告 Panlongcheng yizhi chutu tongqi de weiliang yuansu fenxi baogao, in 盘龙城: 1963-1994年考古发掘报告 *Panlongcheng: 1963-1994 nian kaogu fajue baogao*, 湖北省文物考古研究所 ed. Hubei Institute of Archaeology, Wenwu press, Beijing, pp. 559-573.

Chen, K. and Mei, J., 2006. 山西灵石县旌介村商墓出土铜器的科学分析 Shanxi Lingshixian Jingjiecun Shangmu chutu tongqi de kexue fenxi, in Hai, J. and Han, B. (eds.), 灵石旌介商墓, Science press, Beijing, pp. 209-228.

Chen, K., 2009. 陕西汉中出土商代铜器的科学分析与制作技术研究 *Shaanxi Hanzhong chutu shangdai tongqi de kexue fenxi yu zhizuo jishu yanjiu*, University of Science and Technology, Beijing.

Chernay, I.L., 1980. Pektoral' iz Seletskogo Gorodishcha. *Sovetskaîa arkheologiîa* **1980.4**, 248–251.

Chernikov, S.S., 1951. K Voprosu o Sostave Drevnikh Bronz Kazakhstana. *Sovetskaîa arkheologiîa* **15**,140–161.

Chernykh, E.N. (ed.), 2004. *Kargaly III. Selishche Gornyy: Arkheologicheskiye materialy. Tekhnologiya gorno-metallurgicheskogo proizvodstva Arkheologicheskiye issledovaniya.* Languages of Slavonic Cultures, Moscow.

Chernykh, E.N. and Kuz'minykh, S.V., 1989. *Drevnyaya Metallurgiya Severnoy Evrazii (Seyminskiy-Turbinskiy Fenomen).* Nauka, Moscow.

Chernykh, E.N. and Kuz'minykh, S.V., 1986. Khimicheskiy sostav metalla klada u stantsy Upornaya. *Sovetskaîa arkheologiîa* **1986.3**, 135–138.

Chernykh, E.N., 1963. Izuchenie, sostava mednykh i bronzovykh izdelii metodom spektralnogo analiza. *Sovetskaîa arkheologiîa* **1963.3**, 145-156.

Chernykh, E.N., 1966. Istoriîa drevneĭsheĭ metallurgii Vostochnoĭ Evropy. (History of the oldest metallurgy in Eastern Europe). *Materialy i issledovaniîa po arkheologii SSSR* **132**, 1-145.

Chernykh, E.N., 1966. *Istoriya Drevneyshey Metallurgii Vostochnoy Evropy*. Nauka, Moscow.

Chernykh, E.N., 1966. K Khimicheskoy Kharakteristike Metalla Ingul'skogo Klada. *Sovetskaîa arkheologiîa* **1966.1**, 143–154.

Chernykh, E.N., 1966. Pervye Spektralnye Issledovaniya Medi Dnepro-Donetskoy Kul'tury. *Kratkie Soobshcheniîa Instituta Arkheologii* **106**, 66–68.

Chernykh, E.N., 1967. O Terminakh "Metallurgichsekiy Tsentr", "Ochag Metallurgii" I drugikh. *Sovetskaîa arkheologiîa* **1967.1**, 295–301.

Chernykh, E.N., 1970. *Drevneĭshaîa metallurgiîa Urala i Povolzh'îa*. Nauka, Moscow.

Chernykh, E.N., 1970. O Drevneyshikh Ochagakh Metalloobrabotki Yugo–Zapada SSSR. *Kratkie Soobshcheniîa Instituta Arkheologii* **123**, 23–31.

Chernykh, E.N., 1975. Ai Bunarskiy mednyy rudnik IV tysyacheletiya do N.E. na Balkanakh (issledovaniya 1971, 1972, and 1974 gg.). *Sovetskaîa arkheologiîa* **1975.4**, 132–153.

Chernykh, E.N., 1976. *Drevnîaîa metalloobrabotka na Îugo-Zapade SSSR*. Nauka, Moscow.

Chernykh, E.N., 1978. *Gornoe delo i metallurgiîa v drevneĭsheĭ Bolgarii*. Bulgarian Academy of Sciences Press, Sofia.

Chernykh, E.N., 1980. O Khimicheskom Sostave Metalla Tamanskogo Klada. *Sovetskaîa arkheologiîa* **1980.2**, 150–154.

Chernykh, E.N., 1981. Klad iz Konstantsy i voprosy Balkano-Kavkazskikh svyazey v Epokhu Pozdney Bronzy. *Sovetskaîa arkheologiîa* **1981.1**, 19–26.

Chlenova, N.L., 1972. *Khronologiya Pamyatnikov Karasukskoy Epokhi*. Moscow: Nauka.

Ciugudean, H., Ciuta, M. and Kadar, M., 2006. Consideratii privind caveta piese din depozitul de bronzuri de la Panade (com. Sancel, jud. Alba). *Aplium* **93**, 95–110.

Coghlan, H.H. and Case, H., 1958. Early metallurgy of copper in Ireland and Britain. *Proceedings of the Prehistoric Society* (New Series) **23**, 91–123.

Coghlan, H.H., 1955. Reports of the Ancient Mining and Metallurgy Committee: Analyses of three continental axes and of specimens of Irish ores. *Man* **55**, 6–8.

Coghlan, H.H., Butler, J.R. and Parker, G., 1963. *Ores and metals; a report of the Ancient Mining and Metallurgy Committee*, Royal Anthropological Institute, Occasional paper no. 17, Royal Anthropological Institute of Great Britain and Ireland, London.

Coghlan, H.H., 1964. A metallographic examination of five spearheads from the collections of the Borough Museum, Newbury, Berkshire, England. *Sibrium* **8**, 185–194.

Coghlan, H.H., 1964. An examination of some Bronze Age tools and weapons in the Borough Museum, Newbury, Berkshire, England. *Sibrium* **8**, 161–183.

Coghlan, H.H., 1968. A metallurgical examination of some prehistoric flat and flanged axes. *Archeologia Austriaca* **44,** 61–82.

Collins, W.F., 1931. The corrosion of early Chinese bronzes. *The Journal of the Institute of Metals* **45**, 23-47.

Colpani, F., Angelini, I., Artioli, G. and Tecchiati, U., 2009. Copper smelting activities at the Millan and Gudon Chalcolithic sites (Bolzano, Italy): Chemical and mineralogical investigations of the archaeometallurgical finds, in J.-F. Moreau, R. Auger and A. Herzog (eds.) *Proceedings of 36th International Symposium on Archaeometry, Quebec City, Canada*, pp. 367-373, Laboratoire d'archéologie, Université du Québec à Chicoutimi.

Courcier, A., Lyonnet, B. and Guliyev, F., 2012. Metallurgy during the Middle Chalcolithic Period in the Southern Caucasus: Insight through recent discoveries at Mentesh-Tepe, Azerbeijan, in Jett, P., McCarthy and B., Douglas, J.G. (eds.), *Scientific Research on Asian Metallurgy: Proceedings of the 5th Forbes Symposium at the Freer Gallery of Art*. Archetype Publications, London, pp. 205–224.

Craddock, P.T., 1976. The composition of the copper alloys used by the Greek, Etruscan and Roman civilizations 1. The Greeks before the archaic period. *Journal of Archaeological Science* **3**, 93–113.

Craddock, P.T., 1985. Prehistoric tombs of Ras al-Khaimah. Technical Appendix 1: The composition of the metal artifacts. *Oriens Antiquus* **24**, 97–101.

Craddock, P.T., La Niece, S. and Hook, D.R., 2003. Evidences for the production, trading and refining of copper in the Gulf of Oman during the third millennium BC, in Stöllner, T., Körlin, G., Steffins, G. and Cierny, J. (eds.), *Man and Mining - Mensch Und Bergbau: Studies in Honour of Gerd Weisgerber on Occasion of His 65th Birthday*. Deutsches Bergbau-Museum, Bochum, pp. 103–112.

Cui, J. and Wu, X. 2008. 铅同位素考古研究——以中国云南和越南出土青铜器为例 *Qiantongweisu kaogu yanjiu-- yi Zhongguo Yunan he Yuenan chutu qingtongqi weili*, Wenwu press, Beijing.

Cui, J., Tong, W. and Wu, X. 2012. 垣曲商城出土部分铜炼渣及铜器的铅同位素比值分析研究 Yuanqu shangcheng chutu bufen tonglianzha ji tongqi de qiantongweisu bizhi fenxi yanjiu, 文物 *Wenwu* (7), 80-84.

Cui, J., Wu, X., Tong, W. and Zhang, S. 2009. 山西垣曲商城出土部分铜器的科学研究 Shanxi Yuanqu Shangcheng chutu bufen tongqi de kexue fenxi yanjiu, *Kaogu yu wenwu* (6), 86-90.

Çukur, A. and Kunç, Ş., 1989. Analyses of Tepecik and Tülíntepe metal artifacts. *Anatolian Studies* **39**, 113–120.

Çukur, A. and Kunç, Ş., 1989. Development of bronze production technologies in Anatolia. *Journal of Archaeological Science* **16**, 225–231.

de Marinis, R.C. (ed.), 2006. Aspetti della metallurgia dell'età del Rame dell'antica età del Bronzo nella penisola italiana. *Rivista di Scienze Preistoriche* **56**, 212–272.

de Marinis, R.C., 2005. Évolution et variation de la composition chimique des objets en métal pendant les âges du Cuivre et du Bronze Ancien dans l'Italie septentrionale, in Ambert, P. and Vaquer, J. (eds.), *La première métallurgie en France et dans les pays limitrophes: actes du colloque international, Carcassonne, 28-30 septembre 2002,* Toulouse, Société préhistorique française, pp. 249–264.

de Marinis, R.C., 2006. Circolazione del metallo e dei manufatti nell'età del Bronzo dell'Italia settentrionale, in *Atti della XXXIX Riunione Scientifica, Materie prime e scambi nella preistoria italiana, Firenze, 25-27 novembre 2004.* Istituto italiano di preistoria e protostoria, Firenze, pp. 1289–1317.

De Ryck, I., Adriaens, A. and Adams, F., 2003. Microanalytical metal technology study of ancient near eastern bronzes from Tell Beydar. *Archaeometry* **45**, 579–590.

Degtyareva, A.D. and Kostomarova, Yu.V., 2011. Metall Pozdnogo Bronzovogo Veka lesostepnogo Pritobol'ya. *Vestnik Arkheologii, Antropologii, i Etnografii* **14**, 30–45.

Degtyareva, A.D., 2009. Khimiko-metallurgicheskiye Gruppy Sintashtinskoy Kul'tury. *Vestnik Arkheologii Antropologii i Etnografii* **11**, 29-41.

Degtyareva, A.D., 2010. *Istoriya Metalloproizvodstva Yuzhnogo Zaural'ya v Epokhu Bronzy.* Nauka, Novosibirsk.

Degtyareva, A.D., 2010. Morfologiya i technologiya izgotovleniya ukrasheniy Sintashtinskoy kul'tury. *Vestnik Arkheologii, Antropologii i Etnografii* **12**, 59–70.

Degtyareva, A.D., Grushin, S.P. and Shaykhutdinov, V.M., 2010. Metalloobrabotka naseleniya Eluninskoy cul'tury Verkhney Obi. *Vestnik Arkheologii, Antropologii i Etnografii* **2**, 27–35.

Degtyareva, A.D., Kuzminykh, S.V. and Orlovskaya, L.B., 2001. Metalloproizvodstvo Petrovskikh Plemen (po materialam poseleniya Kulevchi III). *Vestnik Arkheologii, Antropologii i Etnografii* **3**, 23–54.

Demedenko, S.V., 2000. Bronzovye Kotly Rannego Zheleznogo Veka kak Istochnik po Istorii i Kul'ture Drevnikh Plemen Nizhnego Povolzh'ya i Yuzhnogo Priural'ya. Unpublished Dissertation (Kandidat Istoricheskikh Nauk). Moscow.

Desch, C.H., 1938. The bronzes of Luristan. B. Metallurgical Analyses, in Pope, A.U. (ed.), *A Survey of Persian Art from Prehistoric Times to the Present. Volume I.* Oxford University Press, London; New York, p. 278.

Dōno, T., 1965. Chemical investigation of the spearheads and daggers from the Ghalekuti I site, in Engami, N., Fukai, S. and Masuda, S. (eds.), *Dailaman I: The Excavations Ay Ghalekuti and Lasulkan, 1960.* pp. 63, Institute for Oriental Culture, Tokyo.

El Morr, Z. and Mödlinger, M., 2014. Middle Bronze Age artefacts and metallurgical practices at the sites of Tell Arqa, Mougharet el-Hourryieh, Yanouh and Khariji in Lebanon. *Levant* **46**, 28–42.

El Morr, Z. and Pernot, M., 2011. Middle Bronze Age metallurgy in the Levant: evidence from the weapons of Byblos. *Journal of Archaeological Science* **38**, 2613–2624.

El Morr, Z., Cattin, F., Bourgarit, D., Lefrais, Y. and Degryse, P., 2013. Copper quality and provenance in MBI Byblos and Tell Arqa (Lebanon). *Journal of Archaeological Science* **40**, 4291–4305.

Fan, X., and Su, R., 1997. 新干商代大墓铜器合金成分 Xin'gan shangdai damu tongqi hejin chengfen, in 新干商代大墓 *Xin'gan shangdai damu*, eds. Jiangxi Provincial Museum, Jiangxi Institute of Archaeology, and Xin'gan Museum, Wenwu press, Beijing, pp. 241-244.

Feng, F., Wang, Z., Hua, J. and Bai, R., 1982. 殷墟出土商代青铜觚铸造工艺的复原研究 Yinxu chutu Shagdai qingtong gu zhuzao gongyi de fuyuan yanjiu, 考古 *Kaogu* (5), 532-539.

Flek, E.V., 2009. Krestovidnye Podveski Petrovskoy i Alakul'skoy Kultur. *Vestnik Arkheologii, Antropologii, i Etnografii* **9**, 64–71.

Flek, E.V., 2010. Bronzovyye Blyashki Alakul'skoy Kul'tury. *Vestnik Arkheologii, Antropologii i Etnografii* **12**, 87–96.

Fleming, S.J., 2006. The archaeometallurgy of War Kabud, Western Iran. *Iranica Antiqua* **41**, 31–57.

Fleming, S.J., Nash, S.K. and Swann, C.P., 2011. The archaeometallurgy of Period IVB Bronzes at Hasanlu, in de Schauensee, M. (ed.), *Peoples and Crafts in Period IVB at Hasanlu, Iran*. University of Pennsylvania Museum of Archaeology and Anthropology, Philadelphia, pp. 103–134.

Fleming, S.J., Pigott, V.C., Swann, C.P. and Nash, S.K., 2005. Bronze in Luristan. Preliminary analytical evidence from copper/bronze artifacts excavated by the Belgian Mission in Iran. *Iranica Antiqua* **40**, 35–64.

Formigli, E., 1981. Tradizioni e innovazioni della metallotecnica etrusca, in *L'Etruria Mineraria - Atti del XII Convegno di Studi Etruschi e Italici*, Firenze-Populonia-Piombino, pp. 51–78, Istituto di studi etruschi ed italici, L.S. Olschki, Firenze.

Frame, L., 2010. Metallurgical investigations at Godin Tepe, Iran, Part I: the metal finds. *Journal of Archaeological Science* 37, 1700–1715.

Gadizhev, M.G., 1965. O Bronzovykh Bulavkakh Dagestana Epokhu Bronzy. *Sovetskaîa arkheologiîa* **1965.1**, 185–89.

Gailhard, N., 2009. *Transformation du cuivre au Moyen-Orient du Neolithique a la fin du 3eme millenaire: Etude d'une chaine technologique*. BAR International Series 1911.

Gak, E.I. and Yegor'kov, A.N., 2009. Latun' Ergeninskogo Mogil'nika I ee Istorico-metallurgicheskie Kontekst. *Arkheologicheskie Vesti* **16**, 57–62.

Gak, E.I., 2005. Metalloobrabatyvayushchee Proizvodstvo Katakombnykh Plemen Stepnogo Predkavkaz'ya, Nizhnego Dona i Severskogo Dontsa. Unpublished Dissertation (Kandidat Istoricheskie Nauk). Moscow.

Gak. E.I., Mimokhod, R.A. and Kalmykov, A.A., 2012. Sur'ma v Bronzovom Veke Kavkaza i yuga Vostochnoj Evropy. *Arkheologicheskie Vesti* **18**, 174–203.

Gale, N.H., Stos-Gale, Z.A. and Gilmore, G.R., 1985. Alloy types and copper sources of Anatolian copper alloy artifacts. *Anatolian Studies* **35**, 143–173.

Galibin, V.A. 1983. Spektral'nyy analiz nakhodok iz Sumbarskikh mogil'nikov. In Khlopin, I.N., (ed.), *Yugo-Zapadnaya Turkmeniya v Epokhu Pozdney Bronzy*. Nauka, Moscow, pp. 224–234.

Galibin, V.A., 2002. Metall iz Mogil'nik Parkhay II. In Khlopin, I.N. *Epokha Bronzy Yugo-Zapadnogo Turkmenistana*. St Petersburg: Peterburgskoe Vostokovedenie.

Gansu Sheng Wenwu Kagogu Yanjiusuo, 2009. *Chongxin yujiawan zhou mu*. Wenwu Chubanshe, Beijing.

Garagnani, G.L., Imbeni, V. and Martini, C., 1997. Analisi chimiche e microstrutturali di manufatti in rame e bronzo delle terramare. In Bernabò Brea, M., Cardarelli, A. and Cremaschi, M. (eds), *Le Terramare. La Più Antica Civiltà Padana*, pp. 554–566, Electa, Milano.

Garagnani, G.L., Martini, C. and Petitti, P., 1995. Indagini analitico-strutturali su reperti metallici dell'età del rame provenienti dalle necropoli di Selvicciola (Ischia di Castro, VT) e grotta Fichina (Monteromano, VT), in Amico, C.D and Finotti, F. (eds.), *Le Scienze Della Terra E l'Archeometria*, pp. 103–106.

Garagnani, G.L., Spinedi, P. and Baffetti, A., 1993. Caratterizzazione microstrutturale ed analisi chimiche dei reperti metallici, in Baffetti, A., Carancin, G.L. and Conti, A.M. (eds.) *Vulcano a mezzano-Insediamento e produzioni artigianali nella media valle del fiora nell'eta del bronzo*, pp. 87–95, Comune di Valentano.

Geniş, E.Y. and Zimmerman, T., 2014. Early Bronze Age metalwork in Central Anatolia - An archaeometric view from the hamlet. *Praehistorische Zeitschrift* **89**, 280–290.

Gennadiev, A.N., Derzhavin, V.L., Ivanov, V.K., and Smirnov, Yu. A., 1987. O redkoy forme pogrebal'nogo obryada v Predkavkazskoy kul'ture. *Sovetskai͡a arkheologii͡a* **1987.2**, 124—141.

Gettens, R.J., 1969. *The Freer Chinese bronzes, volume 2: technical studies.* Smithsonian Institution, Washington D.C.

Gevorkyan, A.T., 1972. Khimicheskaya kharakteristika metalla iz Lchashenskikh kurganov. *Sovetskai͡a arkheologii͡a* **2**, 172–178.

Giardino, C., Gigante, G. and Ridolfi, S., 2003. Appendix 8.1: Archaeometallurgical studies, in Swiny, S., Rapp, G. and Herscher, E. (eds.), *Sotira Kaminoudhia: An Early Bronze Age Site in Cyprus.* American Schools of Oriental Research.

Giot, P.-R., 1964. Analyses spectrographiques d'objets préhistoriques et antiques, *Travaux du Laboratoire d'anthropologie préhistorique de la Faculté des sciences de Rennes. Laboratoire d'anthropologie préhistorique*, Faculté des sciences, Rennes.

Giot, P.-R., 1975. Analyses spectrographiques d'objets préhistoriques et antiques, *Travaux du Laboratoire d'anthropologie préhistorique de la Faculté des sciences de Rennes. Laboratoire d'anthropologie préhistorique*, Faculté des sciences, Rennes.

Giumlia-Mair, A., Keall, E., Stock, S. and Shugar, A., 2000. Copper-based implements of a newly identified culture in Yemen. *Journal of Cultural Heritage* **1**, 37–43.

Golden, J., Levy, T.E. and Hauptmann, A., 2001. Recent discoveries concerning Chalcolithic metallurgy at Shiqmim, Israel. *Journal of Archaeological Science* **28**, 951-963.

Gorbenko, K.V. and Goshko, T.Yu., 2010. Metalevi virobi z poseleniya Dikiy Sad. *Arkheologiya* **2010.1**: 97–111.

Goryunova, O.I. and Pavlova, L.A., 2008. Metallicheskiye Izdeliya iz Ppogrebeniy Mogil'nika Bronzovogo Veka Kurma XI (oz. Baikal). In *Antropogen Paleoantropologiya, Geoarkheologiya, Etnologiya Azii*, (ed.) Medvedev, G.I., Izd-vo Ottisk, Irkutsk, pp. 53–56.

Goshko, T.Yu. 2011. *Metaloobrobka u naselennya pravoberezhnoy lisostepnovoy Ukrayini za dobi piznoy bronzi.* Institut Arxeologiy NAKU, Kiev.

Goy, J., Le Carlier de Veslud, C., Esposti, Degli, M. and Attaelmann, A.G., 2013. Archaeometallurgical survey in the area of Masafi (Fujairah UAE): preliminary data from an integrated programme of survey excavation and physicochemical analyses. *Proceedings of the Seminar for Arabian Studies* **43,** 127-143.

GRANEP (Grup de Recerca Arqueologica del NordEst Peninsular), 2008. Noves Dades Arqueometallurgiques d'Objectes Metallics de Base Coure Dipositats al Museu d'Arqueologia de Catalunya-Girona. *Cypsela* **17**, 143–48.

Grigor'yev, S., 2015. *Metallurgical production in northern Eurasia in the Bronze Age.* Archaeopress, Oxford.

Grishin, Yu.S., 1971. *Metallicheskiye Izdeliya Sibiri Epokhi Eneolita i Bronzy.* Nauka, Moscow.

Grishin, Yu.S., 1983. The role of copper metallurgy in the Bronze Age and Early Iron of Trans-Baikal, in Konovalov, Ya.B. (ed.), *The Footsteps of Ancient Cultures in the Baikal Region*, Nauka, Novosibirsk, pp. 101-107.

Gunter, A.C. and Jett, P., 1992. *Ancient Iranian Metalwork in the Arthur M. Sackler Gallery and the Freer Gallery of Art.* Arthur M. Sackler Gallery, Washington, DC.

Hakemi, A., 1997. *Shahdad: Archaeological Excavations of a Bronze Age Center in Iran.* IsMEO, Rome.

Halm, L., 1935. Analyse chimique et etude micrographique de quelques objets de metal cuivreux provenant du Tépé-Giyan, in Contenau, G. and Ghirshman, R. (eds.), *Fouilles Du Tépé-Giyan Près de Néhavend, 1931 et 1932*. P. Geuthner, Paris, pp. 135–138.

Halm, L., 1939. Analyse chimique et etude micrographique de quelques pièces de métal et de céramique provenant de Sialk, in Ghirshman, R. (ed.), *Fouilles de Sialk Près de Kashan, 1933,1934,1937*. P. Geuthner, Paris, pp. 205–208.

Hasanova, A., 2014. Influence of Anatolian traditions smelting ancient tin bronze on the development metallurgy III Millennium BC on the Territory of Azerbaijan, in Narimanishvili, G.K., Kvačaże, M., Puturidze, M. and Šanšašvili, N. (eds.), *Problems of Early Metal Age Archaeology of Caucasus and Anatolia*, NM Press, G. Tbilisi, pp. 65–72.

Hasanova, A., 2015. Archaeometallurgical studies metallic artifacts from the Middle Bronze Age sites of Southeast of Azerbeijan. *Global Journal of Human-Social Science* **15**, 1–9.

Hasanova, A., 2016. Archaeometallurgical studies of spearheads and arrowheads of the Middle Bronze Age sites of Azerbeijan. *Mediterranean Archaeology and Archaeometry* **16**, 221–226.

Hauptmann, A. and Pernicka, E., 2004. *Die Metallindustrie Mesopotamiens von den Anfângen bis zum 2. Jahrtausend v.Chr.* Verlag Marie Leidorf., Rahden/Westf.

Hauptmann, A., 2007. *The archaeometallurgy of copper: evidence from Faynan, Jordan*. Springer, Berlin.

Hauptmann, A., Begemann, F. and Schmitt-strecker, S., 1998. Copper objects from Arad: Their compositon and provenance. *Bulletin of the American Schools of Oriental Research* **314**, 1–7.

Hauptmann, A., Begemann, F., Heitkemper, E., Pernicka, E. and Schmitt-strecker, S., 1992. Early copper produced at Feinan, Wadi Araba, Jordan: the composition of ores and copper. *Archaeomaterials* **6**, 1–33.

Hauptmann, A. and Pernicka, E., 2004. *Die Metallindustrie Mesopotamiens von den Anfângen bis zum 2. Jahrtausend v.Chr.* Verlag Marie Leidorf, Rahden/Westf.

Hauptmann, A., Rehren, T. and Schmitt-strecker, S., 2003. Early Bronze Age Copper Metallurgy at Shahr-i Sokhta (Iran), reconsidered, in Stöllner, T., Körlin, G., Steffens, G. and Cierny, J. (eds.), *Man and Mining - Mensch Und Bergbau. Studies in Honour of Gerd Weisgerber on Occasion of His 65th Birthday*. Bergbau-Museum Bochum, Bochum, pp. 197–213.

Hauptmann, A., Schmitt-strecker, S. and Begemann, F., 2015. On Early Bronze Age copper bar ingots from the Southern Levant. *Bulletin of the American Schools of Oriental Research* **373**, 1–24.

Hauptmann, A., Weisgerber, G. and Bachmann, H.G., 1988, Early Copper Metallurgy in Oman, in Maddin, R. (ed.), *The Beginning of the Use of Metals and Alloys. Papers from the Second International Conference on the Beginning of the Use of Metals and Alloys, Zhengzhou, China, 21-26 October 1986*. MIT, pp. 34-51, Cambridge, Mass.

He, T. and Ou, T., 1994. 罗山固始商代青铜器科学分析 Luoshan Gushi shangdai qingtongqi kexue fenxi, 中原文物 *Zhongyuan wenwu* (3), 95-100.

He, T., 1997. 先秦青铜合金技术的初步探讨 Xianqin qingtong hejin jishu de chubu tantao, 自然科学史研究 *Ziran kexueshi yanjiu* **16**(3), 273-286.

He, T., 2001. 盘龙城青铜器合金成分分析 Panlongcheng qingtongqi heji chengfen fenxi, in 盤龍城: 1963-1994, in Hubei Institute of Archaeology (ed.), 年考古發掘報告 *Panlongcheng: 1963-1994 nian kaogu fajue baogao*, Wenwu press, Beijing, pp. 539-544.

Hegde, K.T.M. and Ericson, J.E., 1985. Ancient Indian copper smelting furnaces, in Craddock, P.T. and Hughes, M. (eds.), *Furnaces and Smelting Technology in Antiquity*, pp. 59-70.

Henan Institute of Archaeology and Zhengzhou Institute of Archaeology, 1999. 郑州商代铜器窖藏 *Zhengzhou Shangdai tongqi jiaocang*. Kexue Chubanshe, Beijing.

Henan Institute of Archaeology and Zhengzhou Museum, 1983. 郑州新发现商代窖藏青铜器 Zhengzhou xinfaxian Shangdai jiaocang qingtongqi. *Wenwu* (3), 49-59.

Heskel, D., 1981. *The Development of Pyrotechnology in Iran during the Fourth and Third Millennia B.C.* Harvard University, Harvard.

Hook, D.R., 2007. The Composition and technology of selected Bronze Age and Early Iron Age copper alloy artefacts from Italy, in Bietti Sestieri, A.M. and Macnamara, E. (eds.), *Prehistoric Metal Artefacts from Italy (3500-720BC) in the British Museum*. British Museum, London, pp. 308–323.

Hopp, D., Schaaf, H. and Völcker-Janssen, W., 1992. *Iranische Metallfunde im Museum Altenessen*. Habelt Verlag, Bonn.

Höppner, B., Bartelheim, M., Huijsmans, M., Krause, R., Martinek, K.-P., Pernicka, E. and Schwab, R., 2005. Prehistoric copper production in the Inn valley (austria), and the earliest copper in central Europe. *Archaeometry* **47**, 293–315.

Hsu, Y.-K., 2014. *Chemical Composition of Bronze Artefacts from the British Museum Collection*. Unpublished report prepared for the Department of Asia, British Museum.

Hsu, Y.-K., 2016. *The Dynamic Flow of Copper and Copper Alloys across the Prehistoric Eurasian Steppe from 2000 to 300 BCE*, Unpublished DPhil Thesis, University of Oxford

Hua, J.M. and Li, Z.D., 1982. Shang zhou qingtong rongqi de hejin chenfen – jian lun zhong ding zhi qi de xingcheng. *Xibei gonye daxue xuebao* **2**, 22-40.

Huang, Z., Pan, C., Ni, W. and Chen, G., 2008, 长江中游地区楚墓中出土的青铜箭镞的锈蚀现象及锈蚀机理研究 Changjiang zhongyou diqu chumu zhong chutu de qingtong jianzu de xiushi xianxiang ji xiushi jili yanjiu, 文物保护与考古科学 *Wenwu baohu yu kaogu kexue* **20**(4), 16-25.

Hubei Wenwu Kaogu Yanjiusuo, 2001. *Panlongcheng: 1963-1994 nian kaogu fajue baogao*. Wenwu Chubanshe, Beijing.

IA CASS, 2014. 二里头:1999-2006 (Erlitou: 1999-2006), Wenwu Chubanshe, Beijing.

Inanishvili, G., Some aspects of nonferrous metalworking in the Caucasus–Near East (III–II Millennium BC). In Narimanishvili, G.K., Kvačaže, M., Puturidze, M. and Šanšašvili, N. (eds.), *Problems of Early Metal Age Archaeology of Caucasus and Anatolia*, National Museum Press, Tbilisi, pp. 233–244 (2014).

Isikli, M. and Altunkaynak, G., Some observations on relationships between South Caucasus and North-Eastern Anatolia based on recent archaeometallurgical evidence, in Narimanishvili, G.K., Kvačaže, M., Puturidze, M. and Šanšašvili, N. (eds) *Problems of Early Metal Age Archaeology of Caucasus and Anatolia*, National Museum Press, Tbilisi, pp. 73–93 (2014).

Ji, L., 1997. 河南安阳郭家庄160号墓出土铜器的成分分析研究 Henan Anyang Guojiazhuang M160 hao mu chutu tongqi de chengfen fenxi yanjiu, 考古 *Kaogu* (2), 80-84.

Jia, L.J., 2011. *Qin zaoqi qingtongqi keji kaogu xue yanjiu*. Kexue Chubanshe, Beijing.

Jia, Y. and Su, R.U., 2004. Wuguo qingtong bingqi de jinxiangxue kaocha yu yanjiu. *Wenwu keji yanjiu* **2**, 21-51.

Jia, Y., Liu, P.S. and Huang Y.L., 2012. Anhui nanling chutu bufen qingtongqi yanjiu. *Wenwu baohu yu kaogu kexue* **24**(1), 16-25.

Jin, Z., 2008. 中国铅同位素考古 *Zhongguo qiantongweisu kaogu*. University of Science and Technology Press, Beijing.

Jin, Z., Chase, T., Hirao, Y., Peng, S., Mabuchi, H., Miwa, K. and Zhan, K., 1994, 江西新干大洋洲商墓青铜器的铅同位素比值研究 Jiangxi Xin'gan Dayangzhou shangmu qingtongqi de qiantongweisu bizhi yanjiu, 考古 *Kaogu* **8**(8), 744-747.

Jin, Z., Mabuchi, H., Chase, T., Chen, D., Miwa, K. and Hirao, Y., 1995. 广汉三星堆遗物坑青铜器的铅同位素比值研究 Guanghan Sanxingdui yiwukeng qingtongqi de qiantongweisu bizhi yanjiu, 文物 *W enwu* **2** (2), 80-85.

Jin, Z., Zheng, G., Hirao, Y. and Hayakawa, Y., 2001. 初期中國青銅器の鉛同位素比, in 古代東アジア青銅の流通, *Kodai higashi Ajia seidō no ryūtsū*, eds. Y. Hirao, J. Enomoto and H. Takahama, 鶴山堂 Tyoko, pp. 295-304.

Jin, Z., Zhu, B., Chang, X., Zhang, Q. and Tang, F., 2006. 宝山遗址和城洋部分铜器的铅同位素组成与相关问题 Baoshan yizhi he Chengyang bufen tongqi de qiantongweisu zucheng yu xiangguan wenti, in 城洋青铜器 Zhao, C.(ed.) *Chengyang qingtongqi,* Science press, Beijing, pp. 250-259.

Junghans, S., Sangmeister, E. and Schröder, M., 1960. *Metallanalysen kupferzeitlichen mid frühbronzezeitlichen Bodenfunde aus Europa.* Gebr. Mann, Berlin.

Junghans, S., Sangmeister, E. and Schröder, M., 1968. *Kupfer und Bronze in der frühen Metallzeit Europas, Studien zu den Anfängen der Metallurgie.* Gebr. Mann, Berlin.

Junghans, S., Sangmeister, E. and Schröder, M., 1974. *Kupfer und Bronze in der frühen Metallzeit Europas, Studien zu den Anfängen der Metallurgie.* Gebr. Mann, Berlin.

Kaniuth, K., 2006. *Metallobjekte der Bronzezeit aus Nordbaktrien.* Verlag Philipp von Zabern., Mainz.

Kapitonov, I.N., Lokhov, K.I., Berezhnaya, N.G., Matukov, D.I., Bokovenko, N.A., Zaitseva, G.I., Khavrin, S.V., Chugunov, K.V., and Scott, E.M., 2007. Kompleksnye Izotopnye Issledovaniya Bronzovykh Izdeliy Skifskoy Epokhi iz Razlichnikh Pamyatnikov Tsentral'noy Azii, in *Radiouglerod v Arkheologicheskikh i Paleoklimaticheskikh Issledovaniyakh*, St. Petersburg, IIMK Press, pp. 274–285.

Kashani, P., Sodaie, B. and Zoshk, R.Y., 2013. Arsenical copper production in the Late Chalcolithic period, Central Plateau, Iran. Case study: copper-based artefacts in Meymanatabad. *Interdisciplinia Archaeologica* **IV**, 207–210.

Kashkay, M.A. and Semilkhanov, I.R., 1973. *Iz Istorii Drevney Metallurgii Kavkaza.* ELM Press, Baku.

Kashtanov, L.I. and Kashtanova, M. Ya., 1955. *Khimicheskij Sostav Drevnikh Finskikh Tsvetnykh Metallov.* Trudy Instituta Istorii Estestvoznaniya i Tekhniki VI. IIET, Moscow.

Kashtanov, L.I., 1954. *Khimicheskij Sostav Drevnikh Tsvetnykh Splavov na Territoriya SSSR.* Trudy Instituta Istorii Estestvoznaniya i Tekhniki I. IIET, Moscow.

Kaufman, B., 2013. Copper alloys from the 'Enot Shuni Cemetery and the origins of Bronze metallurgy in the EBIV - MBII Levant. *Archaeometry* **55**, 663–690.

Kaydalov, A.I., 2013. *Gorodishche Ust'-Utyak-1 kak Istochnik po Izucheniyu Kul'turno-Istoricheskikh Protsessov na Territorii Srednego Pritobol'ya v Perekhodnoe Vremya ot Bronzy K Zhelezu i Epokhu Rannego Srednevekov'ya.* Unpublished Dissertation (Kandidat Istoricheskikh Nauk). Kemerovo.

Kenoyer, J.M. and Miller, H.M.L., 1999. Metal technologies of the Indus Valley Tradition in Pakistan and Western India, in Pigott, V.C. (ed.), *The Archaeometallurgy of the Asian Old World.* University of Pennsylvania Museum of Archaeology and Anthropology, Philadelphia, pp. 107–151.

Khalil, L., 1984. Metallurgical analyses of some weapons from Tell el-ʿAjjul. *Levant* **XVI**, 167–170.

Khavrin, S.V., 2000. Tagarskie bronzy, in *Mirovozrenie. Arkheologiya. Ritual. Kul Sbornik statei k 60-letiyu M. L. Podol`skogo.* World of Books Press, St. Petersburg, pp. 183–193.

Khavrin, S.V., 2001. Late Bronze Age metal in the nizhnetëyskoy monuments, in Frojanov, I.Ja. (ed.) *Eurasia through Ages*, Faculty of St. Petersburg State University, St. Petersburg, pp. 117-126.

Khavrin, S.V., 2002. Metallurgiya Sayano-Altaya skifskogo vremeni, in *Ladoga i Severnaya Evraziya ot Baikala do Lamansha. Organizuyushchie puti i svyazyvayushchie tsentry*, Petersburg State University Press, St. Petersburg, pp. 70–71.

Khavrin, S.V., 2003. Metal skifskikh pamyatnikov Tuvy i kurgana Arzhan, in I. Piotrovski (ed.) *Stepi Evrazii v drevnosti i Srednevekov`e*. State Hermitage Press, St. Petersburg, pp. 171–173.

Khavrin, S.V., 2007. Tagarskie Bronzy Shirinskogo Rayona Khakasii, in S.V. Khavrin (ed.) *Sbornik nauchnykh trudov v chest' 60-letiya A.V. Vinogradova*, St. Petersburg, Kul't-Inform-Press, pp. 115–122.

Kienlin, T.L., 2008. *Frühes Metall im nordalpien Raum: eine Untersuchung zu technologischen und kognitiven Aspekten früher Metallurgie anhand der Gefüge frühbronzezeitlicher Beile*. Rudolf Habelt, Bonn.

Knapp, A.B. and Cherry, J.F., 1994. *Provenience Studies and Bronze Age Cyprus: Production, Exchange and Politico-Economic Change*. Prehistory Press, Madison.

Koksharov, S.F., 2012. Pervyy Metall Kondy. *Vestnik Arkheologii, Antropologii, i Etnografiya* **4**, 27–42.

Korenevskiy, S.N., Nakhodka bronzogo topora u g. Pyatigorska, *Sovetskai͡a arkheologii͡a* **1972.3**, 337-338 (1972).

Korenevskiy, S.N., 1976. O Metallicheskikh toporakh severnogo Prichernomor'ya, srednego, I nizhego Povolzh'ya Epokhu Sredney Bronzy. *Sovetskaya Arkheologiya* **1976.4**, 16–31.

Korenevskiy, S.N., 1981. Khimicheskiy Sostav Bronzovykh Izdeliy iz Tliyskogo Mogil'nika. *Sovetskaya Arkheologiya* **1981.3**, 148–162.

Korenevskiy, S.N., 1984. O Metalle Bylymskogo klada. *Kratkie soobshcheni͡a* **177**, 7–19.

Korenevskiy, S.N., 1986. O Metalle Epokhi Bronzy v Gornoj Zone Severo-Vostochnogo Kavkaza. *Sovetskaya Arkheologiya* **1986.3**, 5–15.

Kostomarova, Yu.B. and Flek, E.V., 2008. Metall Khripunovskogo Mogil'nik. *Vestnik Arkheologii, Antropologii, i Etnografii* **33**, 45–54.

Krause, R., 2003. *Studien zur kupfer- und frühbronzezeitlichen Metallurgie zwischen Karpatenbecken und Ostsee, Vorgeschichtliche Forschungen*. Leidorf, Rahden/Westf.

Krizhevskaya, L.Ya. 1977. *Rannebronzovoe Vremya v Yuzhnom Zaural'e*. Leningrad University Press, Leningrad.

Kunç, Ş., 1986. Analysis of Ikiztepe metal artifacts. *Anatolian Studies* **36**, 99–101.

Kuzminykh, S.V. and Chernykh, E.N., 1985. Spektroanaliticheskoe Issledovanie Metalla Bronzogo veka Lesostepnogo Pritobol'ya (Predvaritel'nye Rezultaty). In Potemkin, T.M. (ed.) *Bronzovyy Vek Lesostepnogo Pritobol'ya*, Nauka, Moscow, pp. 346–367.

Kuzminykh, S.V., 1977. O Khimicheskom Sostave Metalla Sokolovskogo Mogil'nika. *Sovetskaya Arkheologiya* **1977.4**, 279–83.

Le Roux, G., Veron, A., Scholz, C. and Doumet-Serhal, C., 2003. Chemical and isotopical analyses on weapons from the Middle Bronze Age in Sidon. *Archaeology and History in Lebanon* **18**, 58–61.

Leese, M.N., Craddock, P.T., Freestone, I.C. and Rothenberg, B., 1986. The composition of ores and metal objects from Timna, Israel, in Vendl, A., Pichler, B., Weber, J., Bonik, G. (eds.), *Wiener Bericht Über Naturwissenschaft in Der Kunst*. Technische Universität, Wien, pp. 90–120.

Leoni, M., 1980. Esame analitico di due asce provenienti dalla zona di Carignano. *Sibrium* **XV**, 47–49.

Levushkina, S. and Flitsiyan, E., 1981. Khimicheskiy Sostav Kinzhala iz Vakhshuvara. *Sovetskaya Arkhaeologiya* **1981.1**, 287–288.

Li, M., 1982. 殷墟金属器物成分的测定报告(一)——妇好墓铜器测定 Yinxu jinshu qiwu chengfen de ceding baogao (yi)-- Fu Hao mu tongqi ceding, 考古学集刊 *Kaoguxue jikan* **2**(2), 181-193.

Li, M., Huang, S. and Ji, L., 1984. 殷墟金属器物成分的测定报告(二)--殷墟西区铜器和铅器测定 Yinxu jinshu qiwu chengfen de ceding baogao (er) Yinxu xiqu tongqi he qianqi ceding, 考古学集刊 *Kaoguxue jikan* **4**, 328-375.

Li, X. and Han, R., 2000. 朱开沟遗址出土铜器的金相学研究 Zhukaigou yizhi chutu tongqi de jinxiangxue yanjiu, in Inner Mongolia Institute of Archaeology and Erdos Museum (ed.), 朱开沟——青铜时代早期遗址发掘报告 *Zhukaigou--Qingtong shidai zaoqi yizhi fajue baogao*, Wenwu press, Beijing.

Li, X., Yun, Y. and Han, R., 2015. Scientific examination of metal objects from the third excavation of Haimenkou site, Western Yunnan. In Srinivasan, S., Ranganathan, S. and Giumlia-Mair, A.R. (eds.) *Metals and Civilizations: Proceedings of the VII International Conference on The beginnings of the use of metals and alloys (BUMA-VII)*, pp. 123-128, National Institute of Advanced Studies, Bangalore.

Li, X.H. and Han, Y.B., 1999. Guoguo mu chutu qingtongqi de caizhi fenxi, in Henan sheng wenwu kaogu yanjiu suo (ed.) *Sammenxia guoguo mu, volume 1*, pp. 539-551.

Liang, H., 2004. 二里头遗址出土铜器的制作技术研究 *Erlitou yizhi chutu tongqi de zhizuo jishu yanjiu*, University of Science and Technology, Beijing.

Liang, S.Q. and Chang, G.N., 1950. 中國古銅的化學成分. Zhonggou gutong de huaxue chengfen (The chemical composition of some early Chinese bronzes). *Zhongguo huaxue huihuizhi (Journal of Chinese Chemistry Society)* **17**, 9-18 ().

Liu, J., 2015. 陕北地区出土商周时期青铜器的科学分析研究——兼论商代晚期晋陕高原与安阳殷墟的文化联系 *Shaanbe diqu chutu shangzhou shiqi qingtongqi de kexue fenxi yanjiu-- jianlun shangdai wanqi Jinshaan gaoyuan yu Anyang de wenhua lianxi,* PhD dissertation, University of Science and Technology, Beijing.

Liu, Y., He, Y. and Xu, G., 2007. M54及M60出土青铜器的成分分析 M54 ji M60 chutu qingtongqi de chengfen fenxi, in IA CASS (ed.), 安阳殷墟花园庄东地商代墓葬 *Anyang Yinxu Huayuanzhuang dongdi shangdai muzang*, Science Press, Beijing, pp. 289-296.

Livadie. C., Paternoster, G. and Rinzivillo, R., 2000, Nota sulle analisi mediante fluorescenza X in riflessione totale di asce provenienti da alcuni nuovi ripostigli del Bronzo Antico della Campania. *Bollettino delle sedute della Accademia Gioenia di Scienze Naturali in Catania.* **33** 5-16. (Data in Le Armi Del Bronzo Antico In Campania: Analisi Con Il Metodo Della Fluorescenza X In Riflessione Totale (Txrf) Di Alcuni Reperti Preistoria E Protostoria In Etruria, Sep 2016, Valentano, Italy (2016).

Lu, L.C. and Hu, Z.S., 1988. *Baoji yu guo mudi*. Wenwu Chubanshe, Beijing.

Luyang Shi Wenwu Gongzuodui, 2002. *Luoyang beiyao xizhou mu*. Wenwu Chubanshe, Beijing.

Lyonnet, B., Akhundov, T., Almamedov, K., Bouquet, A., Courcier, A., Jellilov, B., Huseynov, F., Loute, S., Makharadze, Z. and Reynard, S., 2008. Late Chalcolithic kurgans in Transcaucasia. The cemetery of Soyuq Bulaq (Azerbaijan). *Archäologische Mitteilungen aus Iran und Turan* **40**, 27–44.

Ma, J., 2015. 湖南出土商周青铜器的科学分析与研究 *Hunan chutu shangzhou qingtongqi de kexue fenxi yu yanjiu*, University of Science and Technology China, Hefei.

Ma, J., Jin, Z., Tian, J. and Chen, D., 2012. 三星堆铜器的合金成分和金相研究 Sanxingdui tongqi de hejin chengfen he jinxiang yanjiu, 四川文物 *Sichuan wenwu* **2**, 90-96.

Maddin, R. and Stech-Wheeler, T., 1976. Metallurgical study of seven bar ingots. *Israel Exploration Journal* **26**, 170–173.

Mahboubian, H., 1997. *Art of Ancient Iran, Copper and Bronze: The Houshang Mahboubian family collection.* Philip Wilson, London.

Maksimov, E.K., 1972. Perelyubskiy klad mednykh serpov. *Sovetskaı͡a arkheologiı͡a* **1972.2**, 178–181.

Mao, Z., Peng, Z., Zhang, X. and Peng, J., 1993. 先秦青铜器X射线荧光光谱分析 Xianqin qingtongqi X shexian yingguang guangpu fenxi, 考古学集刊 *Kaoguxue jikan* (13), 303-308.

Markovin, V.L. and Glebov, A.I., 1979. Klad bronzoliteyshchika iz okrestnostey stantsy Akhmetovskoy. *Sovetskaı͡a arkheologiı͡a* **2**, 239–245.

Martynov, A.I. and Bogdanova-Berezovskaya, I.V., 1966. Bronze and bronze casting in the northwestern district of Tagar culture, in Mirzoyeva, V.G., (ed.), *History of Western Siberia*, pp. 66-103, Kemerov Kn. Izd-vo, Kemerovo.

Matteoli, L. and Storti, C., 1974. Metallographic researche on four pure copper flat axes and one related metallic block from an Eneolithic Italian cave. *Historical metallurgy* **16**, 65–69.

Matyushchenko, B.I., 1978. Sredneirtyshskiy Tsentr Proizvodstva Turbinsko–Seyminskikh Bronz, in *Drevnie Kul'tury Altaya I Zapadnogo Sibiri*. Nauka, Novosibirsk, pp. 22–35.

Matyushin, G.N., 1971. Pamyatniki Epokhi rannego metalla Yuzhnogo Zaural'ya. *Kratkie Soobshcheniı͡a Instituta Arkheologii* **127**, 117–125.

Meier, D.M.P., 2008. *Die Metallnadeln von Shahdad – eine funktionstypologische Untersuchung*. Eberhard- Karls Universität Tübingen, Tuebingen.

Meier, D.M.P., 2011. Archaeometallurgical investigations on Bronze Age Metal finds from Shahdad and Tepe Yahya. *Iranian Journal of Archaeological Studies* 1**,** 25-34.

Meliksetian, K. and Pernicka, E., 2010. Geochemical characterisation of Armenian EBA metal artefacts and their relation to copper ores., in Hansen, S., Hauptmann, A., Motzenbäcker, I. and Pernicka, E. (eds.), *Von Maikop Bis Trialeti: Gewinnung Und Verbreitung von Metallen Und Obsidian in Kaukasien Im 4. - 2. Jt v. Chr.* Habelt Verlag, Bonn, pp. 41–58.

Meliksetian, K., Pernicka, E., Avetissyan, P. and Simonyan, H., 2003. Chemical and lead isotope characterisation of Middle Bronze Age bronzes and some Iron Age antimony objects (Armenia). *Archaeometallurgy in Europe* **2**, 311–318.

Merkl, M.B., 2011. *Bell beaker copper use in central Europe: a distinctive tradition?* BAR. Archaeopress, Oxford.

Merpert, N.Ya. and. Munchaev, R.M., 1977. Drevneyshaya Metallurgiya Mesopotamii. *Sovetskaı͡a arkheologiı͡a* **1977.3**, 154–163.

Mı͡arkı͡avichı͡us, A., 1980. Khimicheskiy sostav drevneyshikh bronzovykh izdeliy na territoriya Litvy, in Grigalavichene, Ė. and Mı͡arkı͡avichı͡us, A. (eds.), Drevneĭshie metallicheskie izdelı͡ia v Litve: II-I tysı͡acheletı͡ia do n.ė., pp. 101–105, Mokslas, Vilnius.

Miron, E., 1992. *Axes and Adzes from Canaan*. Franz Steiner Verlag, Stuttgart.

Mongez, A, 1804, Mémoire sur le bronze des anciens et sur une épée antique. *Mémoires de l'Institut National, Classe de Littérature et Beaux-arts* **V**, 187-228.

Moorey, P.R.S. and Schweizer, F., 1972. Copper and copper alloys in Ancient Iraq, Syria and Palestine: Some new analyses. *Archaeometry* **14**, 177–198.

Moorey, P.R.S. and Schweizer, F., 1974. Copper and copper alloys in Ancient Turkey: Some new analyses. *Archaeometry* **16**, 112–115.

Moorey, P.R.S., 1969. Prehistoric copper and bronze metallurgy in Western Iran (with special reference to Lūristān). *Iran* **7**, 131–153.

Moorey, P.R.S., 1971. *Catalogue of the ancient Persian Bronzes in the Ashmolean Museum.* Ashmolean Museum, Oxford.

Moorey, P.R.S., 1985. *Materials and Manufacture in Ancient Mesopotamia: the evidence of archaeology and art: metals and metalwork, glazed materials and glass.* BAR, Oxford.

Moucha, V., 1963. Die Periodisierung der Únĕticer Kultur in Böhmen. *Sborník ČSSA* **3**, 9–60.

Muhly, J.D., 1999. Copper and bronze in Cyprus and the Eastern Mediterranean, in Pigott, V.C. (ed.), *The Archaeometallurgy of the Asian Old World.* University of Pennsylvania Press, Pennsylvania, pp. 15–26.

Muscarella, O., 1974. Decorated bronze beakers from Iran. *American Journal of Archaeology* **78**, 239–254.

Muscarella, O., 1988. *Bronze and Iron: Ancient Near Eastern Artifacts in the Metropolitan Museum of Art.* Metropolitan Museum of Art, New York.

Namdar, D., Segal, I., Goren, Y. and Shalev, S., 2004. Chalcolithic copper artefacts. *Salvage Excavation Reports* **1**, 70–83.

Narimanov, N.G. and Dzahfrov, G.F., 1990. O drevneyshey metalurgii medi na territorii Azerbaydzhana. *Sovetskai͡a arkheologii͡a* **1**, 5–14.

National Tokyo Museum, 2005. *Chūgoku kodai hoppōkei seidōki.* Tokyo: National Tokyo Museum.

Naumnov, D.V. and Minyaev, S.S., 1972. Prilozhenie: Khimicheskiy sostav metallicheskih predmetov Samarskogo Klada. *Kratkie Soobshchenii͡a Instituta Arkheologii* **132**, 89–91.

Neppi Modona, A. (ed.), 1981. *L'Etruria mineraria: atti del XII Convegno di studi etruschi e italici, Firenze-Populonia-Piombino, 16-20 giugno 1979.* L.S. Olschki, Firenze.

Nezafati, N., 2006. *Au-Sn-W-Cu-Mineralization in the Astaneh-Sarband Area, West Central Iran.* Eberhard-Karls University, Tuebingen.

Nezafati, N., Pernicka, E. and Shahmirzadi, S.M., 2008. Evidence on the ancient mining and metallurgy at Tappeh Sialk (Central Iran), in Yalçın, Ü, Özbal, H. and Günhan Paşamehmetoğlu, A. (eds.), *Ancient Mining in Turkey and the Eastern Mediterranean: International Conference AMITEM 2008, June 15-22*, pp. 329-349, Atılım University, Ankara.

Northover, P., 2007. Analysis of non-ferrous metalwork, in Barfield, L.H. (ed.), *Excavations in the Riparo Valtenesi, Manerba, 1976-1994,* Istituto italiano di preistoria e protostoria, Firenze, pp. 292–295.

Novotná, M., 1955, Medené nástroje a problém najstaršej ťažby medi na Slovensku. *Slovenská Archeológia* **3**, 70-95.

Oldeberg, A., 1974. *Die ältere Metallzeit in Schweden I.* Stockholm, Almqvist & Wiksell international 1974-1976. (2 vols.).

Orekhov, P.M., 2006. *Bronzoliteynoe Proizvodstvo Prikam'ya v Postanan'inskiy Period.* Unpublished Dissertation (Kandidat Istoricheskikh Nauk). Izhevsk.

Ottaway, B., 1974. Cluster analysis of impurity patterns in Armorico-British Daggers. *Archaeometry* **16**, 221–231.

Ottaway, B., 1982. *Earliest Copper Artifacts of the Northalpine Region: Their Analysis and Evaluation.* Seminar für Urgeschichte, Bern.

Otto, H. and Witter, W., 1952. *Handbuch der ältesten vorgeschichtlichen Metallurgie in Mitteleuropa.* J.A. Barth, Leipzig.

Oudbashi, O. and Davami, P., 2014. Metallography and microstructure interpretation of some archaeological tin bronze vessels from Iran. *Materials Characterization* **97**, 74–82.

Oudbashi, O. and Hasanpour, A., 2016. Microscopic study on some Iron Age bronze objects from W. Iran. *Heritage Science* **4**, 1–8.

Oudbashi, O., Emami, M.S. and Davami, P., 2012. Bronze in archaeometry: a review of the archaeometry of bronze in Ancient Iran, in Collini, L. (ed.), *Copper Alloys - Early Applications and Current Performance - Enhancing Processes.* InTech, Croatia, pp. 153–178.

Oudbashi, O., Emami, M.S., Malekzadeh, M., Hassanpour, A. and Davami, P., 2013. Archaeometallurgical studies on the bronze vessels from "Sangarashan", Luristan, W-Iran. *Iranica Antiqua* **XLVIII**, 147–174.

Özbal, H. and Earl, B., 1996. EBA tin processing at Kestel/Göltepe, Anatolia. *Archaeometry* **38**, 289–303.

Özbal, H., Adriaens, A. and Earl, B., 1999. Hacinebi metal production and exchange. *Paléorient* **25**, 57–65.

Palmieri, A., Begemann, F., Schmitt-strecker, S. and Hauptmann, A., 2002. Chemical composition and lead isotopy of metal objects from the "Royal" Tomb and other related finds at Arslantepe, Eastern Anatolia. *Paléorient* **28**, 43–69.

Patay, P., Zimmer, K., Szabo, Z. and Sinay, G., 1963. Spektrographische und metallographische untersuching Kupfer- und frühbronzeitlicher fund. *Acta Archaeologica Academiae Scientiarum Hungaricae* **15**, 37-64.

Peltenberg, E., 1982. Early copperwork in Cyprus and the exploitation of picrolite: Evidence from Lemba Archaeological Project, in Muhly, J.D., Maddin, R. and Karageorghis, V. (eds.), *Early Metallurgy in Cyprus, 4000 - 500 BC*, pp. 41–62, The Foundation, Nicosia.

Peng, Z., Liu, Y., Liu, S. and Hua, J., 1999. 赣鄂豫地区商代青铜器和部分铜铅矿料来源的初探 Gan'eyu diqu shangdai qingtongqi he bufen tongqian kuangliao laiyuan de chutan, 自然科学史研究 *Ziran kexueshi yanjiu* **18**(3), 241-249.

Peng, Z., Sun, W., Huang, Y., Zhang, X., Liu, S. and Lu, B., 1997. 赣鄂皖诸地古代矿料去向的初步研究 Gan'ewan zhudi gudai kuangliao quxiang de chubu yanjiu, 考古 *Kaogu* (7), 53-61.

Peng, Z., Wang, Z., Sun, W., Liu, S. and Chen, X., 2001. 盘龙城商代青铜器铅同位素示踪研究 Panlongcheng Shangdai qingtongqi qiantongweisu shizong yanjiu, in Hubei Institute of Archaeology (ed.), 盘龙城1963-1949 年考古发掘报告 *Panlongcheng 1963-1949 nian kaogu fajue baogao*, 湖北省文物考古研究所, Wenwu press, Beijing, pp. 552-558.

Penyak, S.I., 1969. Negrovskiy klad bronzovykh mechey (Zakarpatskaya obl. USSR). *Kratkie Soobshcheni͡ia* **115**, 39–44.

Perucchetti, L., 2008. *Studio Archeometallurgico di Armi del Bronzo Antico provenienti dall'Italia Settentrionale.* Unpublished MSc Thesis, Università degli studi di Milano, facoltà di Scienze Matematiche, Fisiche e Naturali, Milano.

Perucchetti, L., Bray, P., Dolfini, A. and Pollard, A.M., 2015. Physical barriers, cultural connections: prehistoric metallurgy across the Alpine region. *European Journal of Archaeology* **18,** 599–632.

Philip, G., 1991. Tin, arsenic, lead: Alloying practices in Syria-Palestine around 2000 BC. *Levant* **XXIII**, 93–104.

Philip, G., 2003. Copper metallurgy in the Jordan Valley from the third to the first Millennia BC: Chemical, metallographic and lead isotope analyses of artefacts from Pella. *Levant* **35**, 71–100.

Pigott, V.C., 1989. The emergence of iron use at Hasanlu. *Expedition* **31**, 67–79.

Pigott, V.C., Rogers, H.C. and Nash, S.K., 2003. Archeometallurgical investigations at Malyan. The evidence for tin-bronze in the Kaftari phase, in Miller, N.F.and Abdi, K. (Eds.), *Yeki Bud, Yeki Nabud. Essays on the Archeology of Iran in Honor of William M. Sumner*. The Cotsen

Institute of Archaeology, University of California, LA, pp. 161–175.

Pigott, V.C., Rogers, H.C., Nash, S.K., 2003. Archaeometallurgical Invistigations at Tal-e Malyan: Banesh Period Finds from ABC and TUV, in Summer, W.M. (ed.), *Early Urban Life in the Land of Anshan: Excavations at Tal-E Malyan in the Highlands of Iran*. University of Pennsylvania Museum of Archaeology and Anthropology, Pennsylvania, pp. 94–102.

Pike, A., 2002. Appendix: Analysis of Caucasian metalwork – The use of antimonal, arsenical and tin bronze in the Late Bronze Age, in Curtis, J. and Kruszynski, M. (eds), *Ancient Caucasian and Related Material in The British Museum*. British Museum Occasional Paper 121, London, British Museum, pp. 87–98.

Pleslová-Stiková, E., 1985. *Makotřasy: A TRB Site in Bohemia*. Sectio Praehistorica Museum, Nationale, Pragae.

Ponting, M.J., 2013. Scientific analysis of the Late Assyrian Period Copper-alloy metalwork from Nimrud, in Curtis, J. (ed.), *An Examination of Late Assyrian Metalwork: With Special Reference to Nimrud*. Oxbow Books, Oxford.

Pope, J.A., Gettens, R.J. Cahill, J. and Barnard, N., 1967. *The Freer Chinese Bronzes*. Freer Gallery of Art, Smithsonian Institution, Washington D.C.

Potemkina, T.M., 1985. *Bronzoviy Vek Lesostepnogo Pritobol'ya*. Nauka, Moscow.

Potemkina, T.M. and Degtyareva, A.D., 2007. Metall Yamnoy kul'tury Pritobol'ya. *Vestnik Arkheologii, Antropologii, i Etnografii* **8**, 18–39.

Prange, M. and Hauptmann, A., 2001. The chemical composition of bronze objects from Ibri/Selme, in Yule, P. and Weisgerber, G. (eds.), *The Metal Hoard from Ibri/Selme, Sultanate of Oman*. Franz Steiner Verlag, Stuttgart, pp. 75–84.

Prange, M., 2001. 5000 Jahre Kupfer in Oman 2. Vergleichende Untersuchungen des omanischen Kupfers mittels chemischer und isotopischer Analysenmethoden. *Metalla* **8**, 7-126.

Pronin, A.O., 2008. *Bronzoliteynoye proizvodstvov perekhodnoye ot bronzy k zhelezu vremya na yuge Zapadnoy Sibiri*. Unpublished Dissertation (Kandidat Istoricheskikh Nauk), Novosibirsk.

Pyatkin, B.N., 1978. Rezul'taty spektral'nykh analizov bronzovykh predmetov iz mogil'nika Titovo-1. *Drevnie Kul'tury Altaya I Zapadnoy Sibiri*, 63–66 (1978).

Pyatkin, B.N., 1983. Rezul'taty spektral'nogo analiza bronz kurgana Arzhan (The results of spectral analysis of bronzes in Arzhan), in Kiriushin, I.F. (ed.), *Drevniye gornyaki i metallurgi Sibiri (Ancient miners and metallurgists in Siberia)*. ASU Press, Barnaul, pp. 84-96.

Qian, W., 2006. *Xinjiang hami diqu shiqian shiqi tongqi ji qi yu linjin diqu wenhua de guanxi*. Zhishi Chanquan Chubanshe.

Qin, Y., Wang, C.S. and Mao, Z.W., 2001. Liyong ziran kexue shouduan dui wannan tong kuang shi yelian chanwu shuchu fangxiang de chubu yanjiu. *Qingtong wenhua yanjiu* **2**, 124-127.

Rapp, G., 1982. Native copper and the beginning of smelting, in Muhly, J.D., Maddin, R. and Karageorghis, V. (eds.), *Early Metallurgy in Cyprus, 4000 - 500 BC*, pp. 33–40, The Foundation, Nicosia.

Ravich, I.G. and Ryndina, N.V., 1999. Drevnie splavy med`-mysh`yak i problemy ikh ispol`zovaniya v bronzovom veke Severnogo Kavkaza [Ancient alloys of copper-arsenic and problems of their use in the Bronze Age of the North Caucasus]. *Vestnik MGU im. Lomonosova. Seriya: 8. Istoriya*, 1999, no. 4. [Bulletin of the Moscow State University. Series 8. History], 77–98.

Rawson, J., 1990, *Western Zhou Ritual Bronzes from the Arthur M. Sackler Collections: Ancient Chinese bronzes in the Arthur M. Sackler collections*: Arthur M. Sackler Foundation, Washington, D.C.

Riederer, J., 1982. Metallanalysen Nuernberger Statuetten aus der Zeit der Labenwolf-Werkstatt. *Berliner Beiträge zur Archäometrie* **7**, 94–100.

Riederer, J., 1992. Metallanalysen von Luristan-Waffen. *Berliner Beiträge zur Archäeometrie* **11**, 5–12.

Riesch, L.C. and Horton, D., 1937. Appendix I. Technological analyses of objects from Tepe Hissar., in Schmidt, E.F. (ed.), *Excavations at Teppe Hissar Damghan*, pp. 351-361, The University Museum, Philadelphia.

Rittatore Vonwiller, A., Cardini, L. and Cremascoli, F., 1956. La Necropoli di Canegrate. *Sibrium* **3**, 21–36.

Ruzanov, V.D. and Maltaev, K.Zh., 2002. Rezultaty izucheniya Khimicheskogo sostava metalla nakonechnika kop'ya iz Ferganskoy Doliny. *Arkeologicheskie issledovaniya v Uzbekistana* **3**, 167–171.

Ruzanov, V.D., 1988. K voprosu o proiskhozhdenii Chimbaylykskogo klada. *Istoriya Materialnoy Kul'tury Uzbekistana* **22**, 214–219.

Ruzanov, V.D., 1990. Khimicheskiy Sostav Metalla Mogil'nika Chakka. *Istoriya Materialnoy Kul'tury Uzbekistana* **24**, 214–20.

Ruzanov, V.D., 1991. K voprosu ob istochnikakh metall Erkurgana. *Istoriya Materialnoy Kul'tury Uzbekistana* **25**, 211–214.

Ruzanov, V.D., 1996. Rezultaty spektroanaliticheskikh issledovaniy metalla pamyatnikov Tazabagyabskoy kul'tury. *Istoriya Materialnoy Kul'tury Uzbekistana* **27**, 198–202.

Ruzanov, V.D., 2001. Rezultaty Spektro-Analiticheskikh Issledovaniy Medi I Bronz iz Kvartala "Metallistov" Raskopov R-4 i 15 Gorodishcha Erkurgan. *Istoriya Materialnoy Kul'tury Uzbekistana* **32**, 52–56.

Ruzanov, V.D., 2005. O Metalle Mogil'bika Vuadil' Fergany Epokhi Bronzy. In Anorboev, A.A. (ed.), *Istoriya Uzbekistana v Arkheologicheskikh i Pis'mennykh Istochnikakh*, FAN, Tashkent, pp. 277-278.

Ruzanov, V.D., 2006. Novye dannye o toporakh-kel'takh severnogo Uzbekistana. *Iatoriya Materialnoy Kul'tury Uzbekistana* **35**, 309–310.

Ruzanov, V.D., 2008. Nekotorye itogi izucheniya metalla Zamanbabinskoy kul'tury. *Istoriya Materialnoy Kul'tury Uzbekistana* **36**, 47–49.

Ruzanov, V.D., 2010. Nekotorye korrektivy v vopros daty naruchnogo bronzogo brasleta s poseleniya Turtkul'tepa. *Arkeologicheskie issledovaniya v Uzbekistane* **8**, 245–248.

Ruzanov, V.D., 2012. Khimiko-metallurgicheskaya Kharakteristika Metalla Mogil'nika Muminabad. *Istoriya Materialnoy Kul'tury Uzbekistana* **37**, 39–44.

Ruzanov, V.D., 2013. *Metallobrabotka na Yuge Sredney Azii v Epokhu Bronzy.* Institut Arkheologii AN RUz, Samarqand.

Ruzanov, V.D., Anarbaev, A.A. and Reutova, M.A., 2006. Khimiko-metallurgicheskaya kharakteristika metalla Brichmullinskogo klada. *Istoriya Materialnoy Kul'tury Uzbekistana* **35**, 79–82.

Rychner, V. and Kläntschi, N., 1995. *Arsenic, nickel et antimoine: une approche de la métallurgie du bronze moyen et final en Suisse, par l'analyse spectrométrique, Volume 1.* Cahiers d'archéologie romande, Lausanne.

Ryndina, N.V., 1980. Metall v Kul'turakh Shnurovoj Keramika Ukrainskogo Predkarpat'ya, Podonii i Volyni. *Sovetskaîa arkheologiîa* **1980.3**, 24–42.

Ryndina, N.V., Degtyareva, A.D. and Ruzanov, V.D., 1980. Rezultaty Khimiko-tekhnologicheskogo issledovaniya nakhodok iz Shamshinskogo klada. *Sovetskaîa arkheologiîa* **1980.4**, 154–172.

Salvatori, S., Vidale, M., Guida, G. and Gigante, G., 2002. A glimpse on copper and lead metalworking at Altyn-Depe (Turkmenistan) in the 3rd Millennium BC. *Ancient Civilizations* **8**, 69–106.

Sarianidi, V.I., Terekhova, N.N. and Chernykh, E.N., 1977. O ranney metallurgii i metaloobrabotka drevney Baktrii. *Sovetskaîa arkheologiîa* **1977.2**, 35–42.

Sayre, E.V., Joel, E.C., Blackman, M.J., Yener, K.A. and Özbal, H., 2001. Stable isotope studies of Black Sea Anatolian ore sources and related bronze age and Phrygian artefacts from nearby archaeological sites. Appendix: New central taurus data. *Archaeometry* **43**, 77–115.

Selimkhanov, I.R. and Torosyan, R.M., 1960. Metallograficheskij analiz drevnejshikh metallov v Zakavkaz. *Sovetskaya Arkheologiya* **1960.3**, 229–235.

Selimkhanov, I.R., 1960. K issledovaniyu metallicheskikh predmetov iz "Eneolticheskikh" pamjatnikov Azerbajdzhana i Severnogo Kavkaz. *Sovetskaya Arkheologiya* **1960.2**, 89–102.

Selimkhanov, I.R., 1962. Spectral analysis of metal articles from archaeological monuments of the Caucasus. *Proceedings of the Prehistoric Society* **4**, 68–79.

Selimkhanov, I.R., 1964. Was native copper used in Transcaucasia in Eneolithic Times? *Proceedings of the Prehistoric Society (New Series)* **30**, 66–74.

Selimkhanov, I.R., 1966. Ergebnisse von spektralanalytischen Untersuchungen an Metallgegenständen des vierten und dritten Jahrtausends aus Transkaukasien. *Germania* **44**, 221–233.

Selimkhanov, I.R. and Točik, A., 1963. Die Nitra-Gruppe. *Archeologicke Rozhledy* **15**, 716–774.

Sergeyeva, N.F. and Khamzina, E.A., 1975. Bronzovyye izdeliya iz Posol'ska na Baykale. In *Drevnyaya istoriya narodov yuga Vostochnoy Sibiri*, pp. 176–183, IGU Vyp. 3., Irkutsk.

Sergeyeva, N.F., 1981. Drevneyshaya metallurgiya medi yuga Vostochnoy Sibiri (*Ancient copper metallurgy in the south of eastern Siberia)*. Nauka, Novosibirsk.

Shaanxi Kagu Yanjiusuo, 2006. *Huanxian dongyang*. Kexue Chubanshe, Beijing.

Shaanxi Kagu Yanjiusuo, 2012. *Jiangxian hengshui xizhou mudi qingtongqi keji yanjiu*. Kexue Chubanshe, Beijing.

Shalev, S. and Braun, E., 1997. The metal objects from Yiftah'el II, in Braun, E. (ed.), *Yiftah'el: Salvage and Rescue Excavations at a Prehistoric Village in Lower Galilee, Israel*, Israel Authorities Report. pp. 92–96.

Shalev, S. and Northover, P., 1993. The metallurgy of the Nahal Mishmar hoard reconsidered. *Archaeometry* **35**, 35–47.

Shalev, S., 2007. Metallurgical analysis, in Garfinkel, Y. and Cohen, S. (eds.), *The Middle Bronze Age IIA Cemetary at Gesher: Final Report.* American Schools of Oriental Research.

Shalev, S., 2009. Metals and society: Production and distribution of metal weapons in the Levant during the Middle Bronze Age II, in Rosen, A. and Roux, V. (eds.), *Techniques and People: Anthropological Perspectives on Technology in the Archaeology of the Proto-Historic and Early Historic Periods in the Southern Levant.* pp. 69–80, Editions De Boccard.

Shalev, S., Goren, Y., Levy, T.E. and Northover, P., 1992. A Chalcolithic macehead from the Negev, Israel: Technological aspects and cultural implications. *Archaeometry* **34**, 63–71.

Shmygun, P. E, Sergeeva N.F. and Lykhin, J.P., 1981. Pogrebeniya s bronzovym inventarem na severnom Baikale. In *Novoe v arkheologii Zabaikal'ya*, Nauka, Novosibirsk, pp. 46-50.

Simonyan, A.E., 1984. *Kul'tura Epokhi Sredney Bronzy Severnykh Rayonov Armyanskogo Nagor'ya*. Unpublished Dissertation (Kandidat Istoricheskie Nauk), Leningrad.

Song, J. and Nan, P., 2012. *Jiangxian Hengshui Xi Zhou mudi qingtongqi keji yanjiu*. Kexue chubanshe, Bejing.

Sperl, G., 1981. Untersuchungen zur Metallurgie der Etrusker, in *L'Etruria Mineraria - Atti del XII Convegno di Studi Etruschi e Italici, Firenze-Populonia-Piombino*, pp. 29–50, Istituto di studi etruschi ed italici, L.S. Olschki, Firenze.

Starostin, P.N. and Kuzminykh, S.V., 1978. Pogrebenie Liteyshchitsy iz Pyatogo Rozhdestvensokogo Mogil'nika. In *Voprosy Drevney i Srednevekovskoy Arkheologiya Vostochnoy Evropy*. Nauka, Moscow.

Stefan, C.E., 2008. Trei celturi aflate in colectiile Muzeului National de Antichitati. *Materiale si Cerecetari Arheologice (New series)* **4**, 1–9.

Stos-Gale, Z.A., Gale, N.H. and Gilmore, G.R., 1984. Early Bronze Age Trojan metal sources and Anatolians in the Cyclades. *Oxford Journal of Archaeology* **3**, 23–43.

Sulimanov, R.Kh. and Ruzanov, V.D., 1986. Khimiko-Metallurgicheskaya Kharakteristika Medi I Bronz iz Gorodishcha Erkurgan. *Istoriya Materialnoy Kul'tury Uzbekistana* **20**, 197–205.

Sumner, W., 2003. *Early Urban Life in the Land of Anshan*. University of Pennsylvania Museum of Archaeology and Anthropology, Pennsylvania.

Sun, S. and Han, R., 1981. 中国早期铜器的初步研究 Zhongguo zaoqi tongqi de chubu yanjiu, 考古学报 *Kaogu xuebao* **3**(3), 287-302.

Sun, S., Han, R., Chen, T., Zhai, T., Ban, B. and Tian, K., 2001. 盘龙城出土青铜器的铅同位素比测定报告 Panlongcheng chutu qingtongqi de qiantongweisubi ceding baogao, in Hubei Institute of Archaeology (ed.), 湖北省文物考古研究所盘龙城: 1963-1994年考古发掘报告 *Panlongcheng: 1963-1994 nian kaogu fajue baogao*, Wenwu press, Beijing, pp. 545-551.

Sunchugashev, Ya.I., 1957. O Vyplavke Medi v Drevnej Tuve. *Sovetskaya Arkheologiya,***1957.4**, 487–91.

Sunchugashev, Ya.I., 1969. *Gornoye delo i vyplavka metallov v drevney Tuve*. Nauka, Moscow.

Sunchugashev, Ya.I., 1975. *Drevneyshiye rudniki i pamyatniki ranney metallurgii v Khakassko-Minusinskoy kotlovine,* Nauka, Moscow.

Takahama, H., Yasuhiro, H. and Yoshimitsu, H., 2001. 中國北方系民族の青銅器, in Yoshimitsu, H., Junko, E. and Hideki, T. (eds.) *Kodai higashi Ajia seidō no ryūtsū*, 鶴山堂, Tokyo, pp. 187-252.

Tallon, F., Malfoy, J.-M. and Menu, M., 1987, *Métallurgie Susienne I: de la fondation de Suse au XVIIIe avant J.-C.,* Editions de la Réunion des musées nationaux, Ministère de la Culture et de la Communication Paris (1987).

Tatarinov, S.I., 1979. Metalloobrabotka v Epokhu Pozdney Bronzy na Srednem Dontse. *Sovetskaya Arkheologiya* **1979.4** 258–265.

Tavadze, F. and Sakvarelidze, T., 1959. *Bronzy Drevney Gruzii*. ANG SSR Press, Tbilisi.

Thornton, C.P., 2009. *The Chalcolithic and Early Bronze Age Metallurgy of Tepe Hissar, Northeast Iran: A Challenge to the "Leventine Paradigm",* University of Pennsylvania.

Thornton, C.P., Llamberg-Karlovsky, C.C., Liesers, M. and Young, S.M.M., 2002. On pins and needles: tracing the evolution of copper-base alloying at Tepe Yahya, Iran, via ICP-MS analysis of common-place items. *Journal of Archaeological Science* **29**, 1451–1460.

Tian, C., 1981. 从现代实验剖析中国古代青铜铸造的科学成就 Cong xiandai shiyan pouxi Zhongguo gudai qingtong de kexue chengjiu, 机械 *Jixie* (3), 111-126.

Tian, J., 2013. 郑州地区出土二里岗期铜器研究 *Zhengzhou diqu chutu Erligangqi tongqi yanjiu*, University of Science and Technology China, Hefei.

Tian, J., Jin, Z., Li, R., Yan, L. and Cui, J., 2010. An elemental and lead-isotopic study on bronze helmets from royal tomb No.1004 in Yin ruins, *Archaeometry* **52**, 1002-1014.

Tigeyeva, E.V., 2011.Tekhnologiya izgotovleniya metallicheskikh izdeliy Chistolebyazhskogo Mogil'nik. *Vestnik Arkheologii, Antropologii, i Etnografii* **15**, 66–78.

Tigeyeva, E.V., 2013. Khimiko-metallurgicheskaya kharakteristika metalla alakul'skoy kul'tury Srednego Pritobol'ya. *Vestnik arkheologii, antropologii i etnografii* **22**, 31–39.

Tikhonov, B.G. and Grishin, Yu.S., 1960. *Essays on the history of production in the Urals and southern Siberia during the Bronze Age and Early Iron Age*. Nauka, Moscow.

Tishkin, A.A., 2010. *Torevtika v drevnikh i srednevekovykh kul'turakh Yevrazii*. Azbuka, Barnaul.

Todorova, K, Ryndina, N.V. and Chernykh, E.N., 1977. Eneoliticheskiy Metall iz Golyamo Delchevo (Bolgariya). *Sovetskaya Arkheologiya* **1977.1**, 15–26.

Todorova, K., 1981. *Die kupferzeitlichen Äxte und Beile in Bulgarien*. C.H. Beck, Munchen.

Troitskaya, T.N. and Galibin, V.A., 1983. Rezul'taty kolichestvennogo spektral'nogo analiza predmetov epokhi rannego zheleza Novosibirskogo Priob'ya. In *Drevniye gornyaki i metallurgi Sibiri: Mezhvuzovskiy sbornik nauchnykh statey*, pp. 35-47, AGU, Barnaul.

Trufanov, A.A., 2011. Metallicheskie amulety-podveski Severnogo Prichernomor'ya pervykh vekov n. e. *Stratum plus* **4**, 225–270.

Tylecote, R.F. and McKerrell, H., 1986. Metallurgical technology. Note: Examination of copper alloy tools from Tepe Yahya, in Llamberg-Karlovsky, C.C. (ed.), *Excavations at Tepe Yahya, Iran 1967-1975 The Early Periods*, Peabody Museum of Archaeology and Ethnology, Harvard University, Cambridge, Mass.

Tylecote, R.F., 1982. The Late Bronze Age and bronze metallurgy at Enkomi and Kition, in Muhly, J.D., Maddin, R. and Karageorghis, V. (eds.), *Early Metallurgy in Cyprus, 4000 - 500 BC*. pp. 81–99, The Foundation, Nicosia.

Uchida, J., and Iizuka, Y., 2015. 中央研究院收藏殷墟青铜器的冶金学研究 Zhongyang yanjiuyuan shoucang Yinxu qingtongqi de yejinxue yanjiu, in Li Y.-T. (ed.), 纪念殷墟发掘八十周年学术讨论会论文集, Institute of History and Philology, Academia Sinica, Taipei, pp. 81-112.

Université de Rennes, 1984. *Paléométallurgie de la France atlantique: travaux du Laboratoire d'Anthropologie-préhistoire-protohistoire-quaternaire armoricains* Université de Rennes I, Rennes (2 vols).

Vatandoust, A., 1977. *Aspects of Prehistoric Iranian Copper and Bronze Technology*. University of London, London.

Vatandoust, A., 1999. A view on prehistoric Iranian metalworking: Elemental analyses and metallographic examinations, in Hauptmann, A., Pernicka, E., Rehren, T. and Yalçin, Ü. (eds.), *The Beginnings of Metallurgy*, pp. 121–140, Deutsches Bergbau-Museum, Bochum.

Vatandoust, A., Parzinger, H. and Helwing, B., 2011. *Early mining and metallurgy on the western central Iranian plateau: the first five years of work*. Verlag Philipp von Zabern, Mainz.

Vertman. E.G. and Dubova. N.A., 2013. Rekonstruktsiya Xhimicheskogo Sostava Metalla Pamyatnika Bronzogo Veka Gonur Depe (Turkmenistan) po Dannym Analiza Metodom Mass-Spekrometrii s Induktivno Svyazannoy Plazmoy. *Vestnik Tomskogo Gosudarstvennogo Universiteta–Istoriya* **24**, 5–9.

Vigliardi, A., 2002. La grotta del Fontino: una cavità funeraria eneolitica del Grossetano, Millenni (Florence, Italy) ; 4. *Museo fiorentino di preistoria*, Paolo Graziosi, Firenze.

Vinogradov, N.M., Degtyareva, A.D. and Kuz'minykh, S.V., 2013. Metallurgiya I metalloobrabotka v zhizni obitateley ukreplennogo poseleniya Ust'ye 1. *Vestnik arkheologii, antropologii i etnografii* **2013.3**, 4–30.

Vinogradova, N.M., 2004. Zemledel'cheskie pamyatniki yuzhnogo Tadzhikistana v Epoku bronzy. *Arkheologiheskie Vesti* **11**, 72–93.

Wang, Q., 2002. *Metalworking technology and deterioration of Jin bronzes from the TianmaQucun site, Shanxi, China*. Archaeopress, Oxford.

Wang, X., 2013. 山东滕州前掌大墓地出土铜器分析与保护修复 *Shandong Tengzhou Qianzhangda mudi chutu tongqi fenxi yu baohu xiufu*, University of Science and Technology, Beijing.

Webb, J., Frankel, D., Stos, Z.A. and Gale, N.H., 2006. Early Bronze Age metal trade in the Eastern Mediterranean. New compositional and lead isotope evidence from Cyprus. *Oxford Journal of Archaeology* **25**, 261–288.

Weeks, L.R., 1997. Prehistoric Metallurgy at Tell Abraq, U.A.E. *Arabian Archaeology and Epigraphy* **8**, 11 – 85.

Weeks, L.R., 2000. Metal artefacts from the Sharm tomb. *Arabian Archaeology and Epigraphy* **11**, 180 – 198.

Weeks, L.R., 2003. *Early metallurgy of the Persian Gulf: technology, trade, and the Bronze Age World*. Brill, Boston.

Wei, G., 2007. 古代青铜器矿料来源与产地研究的新进展 *Gudai qingtongqi kuangliao laiyuan yu chandi yanjiu de xinjinzhan*, University of Science and Technology China, Hefei.

Wei, G., Qin, Y., Wang, C., Liu B. and Wang, G., 2006. Dajing kuangye yizhi yelian chanwu de shuchu fangxiang. *Bulletin of Mineralogy, Petrology and Geochemistry* **3**, 254–259.

Yagel, O.A., Ben-Yosef, E. and Craddock, P.T., 2016. Late Bronze Age copper production in Timna: new evidence from site 3. *Levant* **48**, 33–51.

Yahalom-Mack, N., Galili, E., Segal, I., Eliyahu-Bhar, A., Boaretto, E., Shilstein, S.and Finkelstein, I., 2014. New Insights into the Levantine copper trade: Analysis of ingots from the Bronze Age and Iron Age in Israel. *Journal of Archaeological Science* **45**, 159–177.

Yalçin, Ü. and Pernicka, E., 1999. Fruhneolithische Metallurgie von Aslkli Hoyuk, in Hauptmann, A., Pernicka, E., Rehren, T. and Yalçin, Ü. (eds.), *The Beginnings of Metallurgy*. Deutsches Bergbau-Museum, Bochum, pp. 45–54.

Yalçin, Ü. and Yalçin, H.G., 2008. Der Hortfund von Tülinepe, Ostanatolien, in Yalcin, Ü. (ed.) *Anatolian Metal IV*. Beihefte Der Anschnitt 21, Bergbau-Museum Bochum, Bochum.

Yang, G. and Ding, J., 1959. 司母戊大鼎的合金成分及其铸造技术的初步研究 Simuwu dading de hejin chengfen jiqi zhuzao jishu de chubu fenxi, 文物 *Wenwu* (12), 27-29.

Yang, J., 2002. 陕西关中地区先周和西周早期铜器的技术分析与比较研究 *Shaanxi guanzhong diqu xianzhou he xizhou zaoqi tongqi de jishu fenxi yu bijiao yanjiu*, University of Science and Technology, Beijing.

Yang, X., 2007. 山西侯马晋侯墓地M9、M91出土青铜残片分析研究 Shanxi Houma jinhou mudi M9/M91 chutu qingtong canpian fenxi yanjiu, 文物科技研究（第五辑),*Wenwu keji yanjiu* (5), 91-101.

Yao, Z., Sun, S., Xiao, L. and Bai, Y., 2005. 成都市博物院几件院藏青铜兵器的分析研究 Chengdushi bowuyuan jijian yuancang qingtong bingqi de fenxi yanjiu, 文物保护与考古科学 *Wenwu baohu yu kaogu kexue* **17**(2), 19-26.

Yegor'kov, A.N., Gak, E.I., and Shishlina, N.I., 2005. Sostav Metalla Kal'mikii v Bronzovom Veke. *Arkheologicheskie Vesti* **12**, 71–76.

Yener, K.A., Sayre, E.V., Joel, E.C., Özbal, H., Barnes, I.L.and Brill, R.H., 1991. Stable isotope studies of Central Taurus ore sources and related artefacts from Eastern Mediterranean Chalcolithic and Bronze Age sites. *Journal of Archaeological Science* **18**, 541–577.

Youshimitsu, H., (ed.) 2001. *Kodai higashi Ajia seido no ryutsu*. Kakuzando Shuppanbu, Tokyo.

Yu, Y., 2015. *Hubei Suizhou Yejiashan mudi chutu Xi Zhou qingtongqi de kexue fenxi yanjiu*. PhD dissertation, University of Science and Technology, Beijing.

Yule, P., Weisgerber, G., Prange, M. and Hauptmann, A., 2001. *The metal hoard from Ibri/Selme Sultanate of Oman*. Franz Steiner, Stuttgart.

Zanini, A., 2002. I manufatti in rame del Fontino, in Vigliardi, A. (ed.*), La grotta del Fontino: una cavità funeraria eneolitica del Grossetano,* pp. 189–202, Museo fiorentino di preistoria "Paolo Graziosi", Firenze.

Zhang, X.M., Liu, Y. and Zhou B.Z., 1999. Zhouyuan yizhi ji yu guo mudi chutu qingtongqi xiushi yanjiu. *Wenwu baohu yu kaogu kexue* **11**(2), 7-18.

Zhao, C., 2004. 安阳殷墟出土青铜器的化学成分分析与研究 Anyang Yinxu chutu qingtongqi de huaxue chengfen fenxi yu yanjiu, 考古学集刊 *Kaoguxue jikan* **15**(15), 243-268.

Zhao, C., 2005. 前掌大墓地出土铜器的化学组成分析与研究 Qianzhangda mudi chutu tongqi de huaxue zucheng fenxi yu yanjiu, 滕州前掌大墓地 *Tengzhou qianzhangda mudi*, 中国社会科学院考古研究所, Wenwu press, Bejing.

Zhao, C., Du, J., Xu, H., Sun, S. and Liang, H., 2006. 河南偃师二里头出土部分铜器的化学组成分析 *Henan yanshi erlitou chutu bufen tongqi de huaxue zucheng fenxi*, in 文物保護與科技考古Wenwu baohu yu keji kaogu, 34-36, 西北大學文博學院, 中國化學會應化委員會考古與文物保護化學委員會, and 中國科技考古學會, Northwest University, Conservation Science Panel of Chinese Chemistry Society, Committee of Archaeological Science in China eds., Sanqin press, Xi'an.

Zhao, C., Yue, Z. and Xu, G., 2008. 安阳殷墟刘家庄北1046号墓出土铜器的化学组成分析 Anyang Liujiazhuangbei 1047 hao mu chutu tongqi de huaxue zucheng fenxi. *Wenwu* (1), 92-94.

Zhongguo Shehui Kexueyuan Kaogu Yanjiusuo, 2006. *Xujianian siwa wenhua mudi*. Kexue Chubanshe, Beijing.

Zhou, W., Chen, J., Lei, X., Xu, T., Chong, J. and Wang, Z., 2009. Three western Zhou bronze foundry sites in the Zhouyuan area, Shaanx province, China, in Mei, J. and Rehren, T. (eds.), *Metallurgy and civilisation: Eurasia and beyond,* Archetype, London, pp. 62-71.

Zhu, F., 1995. 古代中国青铜器 *Gudai Zhongguo qingtongqi*, University of Nankai Press.

Zhuravlev, A.P. and Vrublevskaya, E.L., 1978. Ranniy Etap Metalloobrabotki v Karelii. *Sovetskai͡a arkheologii͡a* **1978.1**, 154–165

Zimmerman, T. and Yildirim, T., 2007. Land of Plenty? New archaeometric insights into Central Anatolian Early Bronze Age metal consumption in funeral contexts. *Antiquity* **81**(314) Project Gallery.

Zimmerman, T., 2007. Anatolia as a bridge from north to south? Recent research in the Hatti heartland. *Anatolian Studies* **57**, 65–75.

Zyablin, L.P., 1977. *Karasukskiy mogil'nik Malyye Kopony III*. Nauka, Moscow.

Index

A

B

C

M

N

O

P

Q

R